ISBN 978-1-332-04385-9
PIBN 10274924

Typical Modern Microscope.—Made by W. Watson and Sons, Ltd., to t
Specification of the late Dr. Henri Van Heurck, Antwerp, for Pho
Micrographic and High-Power Work.

MODERN MICROSCOPY

A Handbook for Beginners and Students

BY

M. I. CROSS

AND

MARTIN J. COLE

LECTURER IN HISTOLOGY AT COOKE'S SCHOOL OF ANATOMY

FOURTH EDITION

REVISED AND ENLARGED

WITH CHAPTERS ON SPECIAL SUBJECTS BY VARIOUS WRITERS

CHICAGO

CHICAGO MEDICAL BOOK COMPANY

1912

PREFACE TO FOURTH EDITION

In the preparation of this new edition, the original intention that it should be for beginners and students has been steadily kept in view. Numerous additions and revisions have been made in order to bring the book into line with present-day knowledge and methods.

Through the kindness of specialists in their own departments, a third part has been added, giving information on many subjects in which amateur microscopists in particular are interested. To these contributors, and to numerous other friends who have assisted by advice, suggestions, and correcting proofs, thanks are most cordially given.

M. I. C.

London,
November, **1911**.

PREFACE TO FIRST EDITION

THIS handbook is not intended to be an exhaustive treatise on the microscope, nor to give particulars of the various patterns of instruments that are made, of which details can be seen in the makers' catalogues, but to afford such information and advice as will assist the novice in choosing his microscope and accessories, and direct him in his initial acquaintance with the way to use it.

The directions for preparing microscopic objects by Mr. Martin J. Cole are the outcome of a very long experience as a preparer of Microscopic Objects of the highest class, and cannot fail to be of the greatest service to the working microscopist.

M. I. CROSS.

CONTENTS

PART I

CHAPTER I

THE MICROSCOPE-STAND

CHAPTER II

OPTICAL CONSTRUCTION

CHAPTER III

ILLUMINATION AND ILLUMINATING APPARATUS

CONTENTS

CONTENTS

CHAPTER IX

CHAPTER X

CHAPTER XI

CHAPTER XII

CHAPTER XIII

CHAPTER XIV

THE PREPARATION OF VEGETABLE TISSUES FOR MOUNTING IN GLYCERINE JELLY, ACETATE OF COPPER SOLUTION, ETC.

CHAPTER XV

CUTTING, GRINDING, AND MOUNTING SECTIONS OF HARD TISSUES— PREPARING METAL SPECIMENS

CHAPTER XVI

PREPARING AND MOUNTING ENTOMOLOGICAL SPECIMENS FOR THE MICROSCOPE

CHAPTER XVII

CRYSTALS AND POLARISCOPE OBJECTS

CHAPTER XVIII

CLEANING AND MOUNTING DIATOMS, POLYCYSTINA, AND FORAMINIFERA

PAGES

CHAPTER XIX

DRY MOUNTS

CHAPTER XX

FINISHING OFF SLIDES

PART III

CHAPTER XXI

AN INTRODUCTION TO THE USE OF THE PETROLOGICAL MICROSCOPE, BY FREDERIC J. CHESHIRE, F.R.M.S.

CHAPTER XXII

ROTIFERA, BY C. F. ROUSSELET, F.R.M.S.

LIST OF ILLUSTRATIONS

MODERN MICROSCOPY

INTRODUCTION

WE have in this book to do with the microscope of to-day, and the history of its development, interesting as it must be, cannot be traced here. Since the first edition of this book was written, eighteen years ago, the microscope has become increasingly a necessity of civilized life. The vast number of discoveries connected with bacteriology, diseases of the blood and tissues, and almost everything that affects our well-being, have been, and are, due to investigations and observations which are made by the aid of the microscope. There is, consequently, a rapidly increasing class of professional workers ; and whereas in days gone by the amateur was the principal user of the microscope, he has been far outstripped by those to whom the instrument is a vital necessity. Hundreds of microscopes are now manufactured to the one of years ago. Naturally, instruments have been designed to suit the special needs of the army of workers, and the old order of things has been changed. With the increasing demand, keen emulation amongst manufacturers has resulted in improvements in both optical and mechanical arrangements, and the so-called students' series of objectives by the leading makers of to-day, at nominal prices, are in many instances superior to the most expensive lenses of former days. Those who are interested in the scientific and artistic aspects of microscopy can view with nothing but satisfaction the distinctly progressive nature of everything connected with the instrument itself and the absorbing secrets it reveals. That the onward march will continue is certain, for the issues which depend on the microscope for their solution are ever increasing in number. The demands which are made on it are constantly growing in exactness and variety, and there are always ready minds and willing hands to devise facilities for meeting them. While it is true that the increasing use of the microscope has been an important element in the improvements that have taken place, the fact cannot be overlooked that much has been due to the interest, unceasing criticism, impartial examination, and well-merited recommendation, together with the suggestion fraught of knowledge, skill, and thoughtful consideration on the part of expert amateurs, many of whose names are familiar to every worker with the microscope.

To the microscopical societies also the evolution of the microscope is due in no small degree, and especially does this apply to the Royal Microscopical Society and the Quekett Microscopical Club, both of which meet in London at 20, Hanover Square. Every improvement in the instrument and its accessories that takes place is presented to these societies for criticism, and in connection with both of them, as officers and members, are men who have attained the highest eminence in microscopical science and manipulation, whose judgments have influenced and moulded the character of microscopy, and who are ever willing to assist by advice and suggestion any who will avail themselves of their experience. It is most desirable that microscopists should become members of a good microscopical society, and those just mentioned enjoy the highest position in England.

It is surprising that, notwithstanding the pleasures and advantages that are associated with microscopical work, microscopy has not the hold on people of refined tastes that its merits should fairly claim for it. It does not seem to be realized that the microscope will unfold its wonders and beauties without that long and careful study which is the necessary preliminary to the majority of scientific pursuits. Those who may be induced to use a microscope in the first place for pleasure and recreation will quickly find their inclinations leading them to a desire for fuller knowledge concerning the subjects which may come under their notice, and by degrees this instrument will become the means of the acquirement of a very liberal education, for its influence is not merely confined to one kingdom; it embraces every tangible and intangible subject, whether it be the air we breathe, with its myriads of invisible friends and foes to human well-being, or the floors of oceans, with their minute flinty shells bearing markings which exceed in accuracy the power of any draughtsman to depict, and which in themselves are invisible to the naked eye, many of them measuring but the $\frac{1}{1000}$ of an inch.

Astronomy, with all the wonders that are associated with the study, demands many a night vigil, an expensive instrument, and a suitable observatory, and even then there is always a sense of dissatisfaction when accounts are read of observations of fellow-workers who are more highly favoured in that they possess glasses of very large aperture, which the average astronomical student could not aspire to. How dependent the observer is on the weather, too!

Photography, with its manifold uses and the pleasant memories associated with numerous pictures that are secured, cannot be compared with the microscope, more especially for the long winter evenings.

For any kind of recreation to produce the mental rest which is required by the man of business, a fresh set of faculties must be brought into play, and no better method can be imagined for the purpose than the introduction to a world whose variety is surprising and illimitable, whose form is lovely and unique, and whose subjects can never be met with excepting in the quiet observations through the microscope-tube in the study.

Perhaps some people may hesitate to attempt working with the microscope, not caring to use it merely as a means of amusement, and mistrusting their ability to employ it scientifically. They reflect that every department has its untiring, experienced workers, and available ground appears to have been gone over so repeatedly that it would seem hopeless for an amateur to attempt to add to existing knowledge on any subject. This idea is a mistaken one, and any microscopist who uses his instrument thoughtfully will be surprised at the manner in which the love for the work will grow upon him, and how gradually he will become master of some special department which he has adopted as his own. On this point we would echo the words of a well-known microscopist: 'It needs no marvellous intellect, no special brilliancy, to succeed in a scientific study; work at it ardently and perseveringly, and success will follow.'

In order that the best results may be obtained, there must be a correct understanding of the general principles on which the instrument should be worked, and the equipment so selected that each part is in tune with the other, and by co-operative working in skilful and interested hands the whole may give the maximum effect. It is no uncommon thing to find work ill-done through insufficient appreciation of elementary facts; it is used too much as a tool, and too little as an instrument. This applies particularly to the professional worker, who frequently has routine work to do, and does not concern himself with half the possibilities of his microscope, so long as he is able to do what is immediately necessary in the most rapid manner. The worker who is wisely equipped, and brings to his study the intelligence that is necessary, will cause his microscope to reveal to him the best it is capable of, and will find a fresh delight in each new structure that is unfolded in all its striking detail to him.

In the succeeding pages the use and proper place of both the microscope and its accessories are indicated in the plainest manner, and if the rules of manipulation which are given are followed, success is insured.

PART I

CHAPTER I

THE MICROSCOPE-STAND

As one looks through the catalogues of the various dealers, and notices the microscope-stands varying in price from £2 to £40, a feeling of bewilderment arises as to what is essential and what can be dispensed with. We will, then, examine the parts, describe their uses and advantages, and state what is necessary for a beginner.

Here let us advise intending purchasers not to buy a microscope unless it bear the name of a well-known manufacturer : a good workman is never ashamed of his handiwork. There are many very inferior instruments that look tempting, but a practical acquaintance with them soon discovers their weak points and inefficiency. If the user is at all progressive, an instrument of inferior quality is either speedily discarded in favour of a well-made one, or it may, on the other hand, cause him to become disheartened, and attribute want of success to his own incapacity instead of the poor quality of the instrument.

Although a good second-hand instrument may be occasionally met with, great discretion is required in purchasing, because improvements may have been introduced since its manufacture, or some damage may have occurred to the optical parts. If it be obtained from a respectable dealer who understands his business, and will give a guarantee of condition, there is some inducement ; but a friend who is up to date in microscopy is generally the best to advise. In all cases before purchasing, a catalogue should be obtained from the maker whose name the instrument bears, so that it may be ascertained whether the pattern is still made, or is antiquated and out of date. It is much better to buy a good stand, capable and worthy of receiving additional apparatus from time to time, rather than an inferior instrument that is completely furnished with objectives and accessories. These latter rarely engender pride of ownership, and are often relegated to some obscure corner after a short acquaintance; whereas, if a good instrument be purchased, with but one objective to start with, there is always a pleasure in working with it, and a peculiar fascination from its quality—a satisfaction in feeling that one has something superior.

A microscope-stand of large size is frequently referred to as ' powerful '; however, the magnifying power does not depend on this quality, but on the optical parts—the eyepiece and objectives which are used. The tendency in modern instruments has been to minimize the size, for the reason, no doubt, that workers in laboratories use them in the upright position, and inconvenience in looking down the tube is reduced, if the microscope is as short as possible. This has unfortunately led to such contracted proportions that efficient working is considerably interfered with. It is, therefore, important that the microscope that is chosen shall have sufficiency of height for all apparatus to be freely used with it.

On p. 7 a typical student's microscope is figured, by reference to which the different parts of the instrument will be made clear.

Fig. 1.—A is the stand, or foot.

B is the tailpiece carrying the mirror (C), with which light is thrown upon the object.

D is the under-fitting, into which are fitted the sub-stage condenser, polarizer, etc.

E is the stage on which the object is placed.

F is the limb carrying the body (G).

At the lower end of the body is a nosepiece (H), having a screw into which the objective is fitted.

At the upper end of the body is a sliding fitting called the draw-tube (J), by means of which additional magnification may be obtained, and into this draw-tube the eyepiece, or ocular (K), fits.

L is a rackwork, by means of which the body (G) is raised

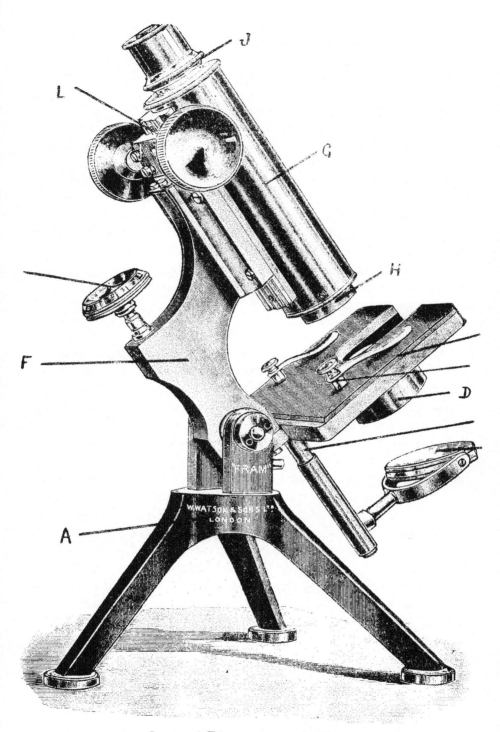

Fig. 1.—A Typical Student's Microscope.

The 'Fram' Microscope.   By W. Watson and Sons, Ltd.

and lowered in order to focus the objective upon the object
which is placed on the stage (E).

M is the milled head controlling the fine adjustment, which
imparts a delicate motion to the body, in order that the objective
may be more exactly adjusted than would be possible with the
rackwork (L) when using high magnifying power.

O is one of the springs with which the object is held in
position.

We have selected the instrument (Fig. 1) because, from practical acquaintance with it, we are able to strongly recommend it
for a beginner's microscope, worthy of receiving additions from
time to time as means may permit. Still, it should only be
considered as a typical one.

The Foot.

Since the first edition of this book was issued, a decided
reaction has taken place in regard to the form of foot on which
the microscope is mounted. There has scarcely been a modern
writer of repute who has not urged the necessity and importance
of having such a mounting for the microscope as shall secure
for it absolute rigidity, whether it be used vertically, inclined,
or horizontally for photography. No foot so fully answers these
requirements as the tripod pattern. Rarely does it happen that
the bench or table on which work is done is absolutely level,
and the tripod is the only pattern that naturally adjusts itself to
such inequalities of surface. It cannot, therefore, be too emphatic-
ally insisted that microscopists should select this pattern in
preference to any other. At first sight this feature may appear
to be a somewhat trivial one, and especially so to a novice ; yet
minor details have a marked significance in the satisfactory
execution of his work. It will be found advantageous to have the
foot shod with cork, as thereby the microscope is in a degree
insulated from vibration, and the risk of scratching the surface
of the table on which it is being used is avoided. It must be
clearly understood, however, that even this form of foot must
be made in correct proportion, or its advantages will be
minimized.

The Jackson model foot, which was suited particularly to

instruments of large size, has not the popularity that it once enjoyed. It is, nevertheless, in point of rigidity and convenient

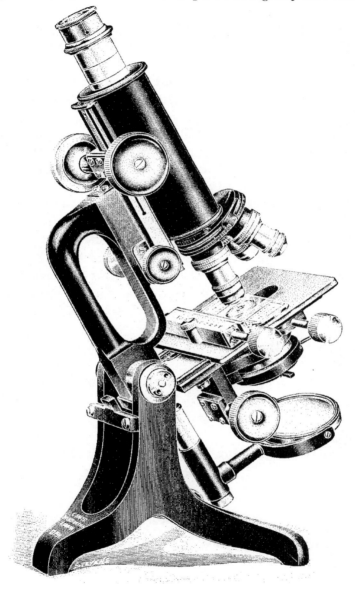

FIG. 2.—D.P.H. MICROSCOPE, BY C. BAKER, SHOWING JACKSON FORM OF FOOT.

shape, one of the best. It will be seen on the instrument in Fig. 2 above.

The type of foot known as the 'horseshoe' is preferred by numerous workers all over the world—we refer particularly to medical students and laboratory workers. To them it is quite satisfactory, because, using the microscope in the upright position, it is sufficiently stable, and allows convenient access to the mirror and sub-stage condenser. It is also very compact, and if unconsciously it be brought over the front edge of the bench, it will remain firm, where the tripod instrument will fall completely over.

The amateur, however, generally uses his microscope inclined at an angle, and in this position the instrument with the horseshoe foot is invariably top-heavy, and has a tendency to fall over with the least knock or pressure.

A modification of the horseshoe foot has of recent years been introduced, and finds much favour. It has all the advantages of the horseshoe, and, if properly proportioned, many of the disadvantages are overcome. The proper proportioning is therefore of great importance ; this is too frequently overlooked.

All considered, the tripod foot has so many points of advantage that it is unhesitatingly to be recommended in preference to any other.

The selection of the foot of the microscope would therefore be in the following order :

1. The tripod foot, as fitted to the instrument illustrated on p. 7.

2. Jackson form of foot, as fitted to the instrument illustrated on p. 9.

3. The modified horseshoe foot, which will be seen on p. 23.

4. The Continental or horseshoe foot, as shown on p. 17.

The Stage.

The stage of the microscope on which the object is placed for examination may be divided into two classes: (1) mechanical, and (2) plain.

THE MECHANICAL STAGE.—The instruments figured on p. 25. and frontispiece are provided with this type of stage, in which by the turning of two milled heads which are attached to screws, plates are moved in dovetailed grooves one over the other, in rectangular directions, carrying the object with them. A first-class microscope should be provided with this form of stage ; in fact, there is no means so suitable for systematically examining an object as is afforded by it. In addition to these mechanical movements, if a bar be fitted to slide in a vertical direction on the top plate the efficiency of the stage will be greatly increased. The mechanical stage lends itself to the adaptation of further important movements. A means of rotating the object is an essential in most classes of work. For this purpose the lowest plate of the stage is usually fitted to rotate on the fixed centre of the base-plate, the mechanical movements acting above it. It is then termed a concentric rotating stage, the object remaining in the field during the whole rotation of the stage. In mechanical stages of economical construction the rotating plate is occasionally fitted above the mechanical movements, and is carried by them, in which case it does not rotate concentrically. The object can, notwithstanding, be kept in the centre of the field by constantly re-setting it with the mechanical screws during the rotation of the plate. Some stages of the concentric form are arranged to rotate by rackwork and pinion ; although this is not really an essential, it is often convenient ; it also prevents the stage from rotating accidentally, especially while photographing. When it is provided, it should have the pinion-wheel so arranged that it may be disengaged from the rack and replaced instantly.

Centring screws to the concentric rotating stage, by means of which the axis of the stage may be made true with any objective, will be found a useful addition, especially if petrological work is to be done. Divisions to the periphery of the stage for reading the angle through which the stage is rotated are not advantageous for ordinary purposes, but for chemical and petrological work they are a necessity.

A remark is necessary respecting the free opening in the centre of the mechanical stage. It is essential that the stage plates, when carried by the mechanical movements, may in no position be obstructed by the top of the sub-stage condenser. Many of the sub-stage condensers in use are of somewhat large diameter, and, unless this feature has received proper thought and arrangement, constant difficulty and trouble will arise.

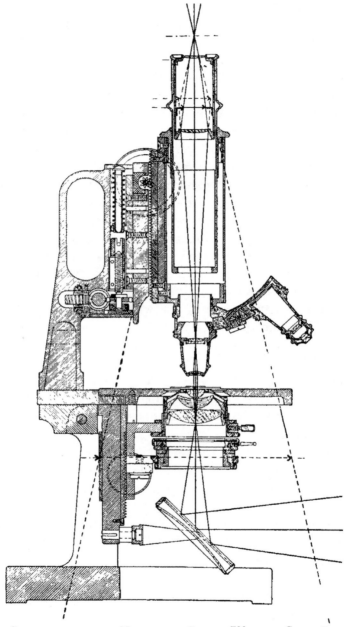

Fig. 3.—Section through Microscope-Stand Ill., by Carl Zeiss, with
Diagram of Path of Rays with Low-Power Objective in Use, showing
also Construction of Fine Adjustment.

ATTACHABLE MECHANICAL STAGES.—In recent years a variety of mechanical stages, which can be attached to or removed from an ordinary plain stage microscope, have been introduced. Some of these possess merit, but, taken as a whole, they are inaccurate in working, and at their best are not for one moment to be compared with the mechanical stage, which has been built as an integral part of the microscope, and no microscopist who wishes to do himself and his work full justice should use such a fitting. They frequently fail to act well when an immersion objective is used, especially if the oil is thick ; and are impossible with an oil immersion condenser, because the oil on the under surface of the object slip is drawn along the stage plate, and sucks up all oil by capillary attraction.

For some reason, Continental manufacturers rarely fit their instruments with the mechanical stage as it is understood in England, but have always recommended and arranged for the adaptation of an attachable stage. Where mechanical movements are found to be an essential for certain work by the possessor of an instrument with a plain stage, and from reasons of inconvenience or impracticability the plain stage cannot be exchanged for a proper mechanical one, then, and then only, should the attachable form be resorted to.

For many classes of work, a stage with a very long range of movement is found advantageous. Many ingenious devices to effect this have been invented, but in nearly every case they present drawbacks over stages of more limited movement, and it is for the consideration of the worker whether the convenience of the long range is of higher importance to him than any slight disadvantage which he may have to suffer.

For this reason the attachable mechanical stages are particularly appreciated, the self-contained method of constructing the horizontal movement allowing of extra travel being given to the object-carrier. By this means the object can usually be carried about 2 inches.

Several long-range mechanical stages of the fixed type are offered by opticians. Messrs. Swift and Son and W. Watson and Sons both have excellent designs, the former in their I.M.S. microscope, and the latter in their Scop and Bactil stages, all of which give 2 inches of horizontal movement. Usually, however, for the work of the amateur, mechanical movements of 1 inch (25 millimetres) in both directions are found ample.

FINDERS TO MECHANICAL STAGES.—Divided scales, reading to parts of an inch or millimetre, fitted to the plates of the mechanical stage, will be found of great utility. By means of such an arrangement, important parts of an object can be noted and subsequently refound. For instance, supposing a specimen were being examined, and an important feature were observed to which future reference would be desirable, it would only be needful to take the reading of the divisions on the stage, and record them on the slide—say, horizontal, 24 ; vertical, 20. On future occasions, on setting the stage readings at the same points and placing the object in the same position on the stage (for which purpose nearly all mechanical stages have a stop-pin, against which the slide can be set), the special feature would be at once in the field of view. These divisions can also be used for roughly measuring objects, the *modus operandi* of which is given in the instructions for the measurement of objects, p. 123.

If a mechanical stage is selected, it should be a good one, for if badly made it is far less advantageous than a plain stage ; also the frictional parts should be sprung, and fitted with adjusting screws, so that compensation may be made for wear and tear.

PLAIN STAGES.—The stage of the microscope shown on p. 7 has two flat springs only, to hold the object in position on the surface, and the movement of the object is effected by the fingers. For cursory examinations this answers every purpose ; but where systematic work is to be done something more is needed, and this, when a mechanical stage is not provided, should take the form of a bar reaching completely across the stage and sliding in a vertical direction. If properly fitted and sprung, it will travel freely when gently pressed with one hand only. The object is carried by it, and can be moved in a horizontal direc-tion upon this bar. With a little practice the fingers become educated to the work, enabling examinations to be conducted with the highest powers almost as rapidly and systematically as with the mechanical screws. The sliding-bar should further be provided with two flat springs, so fitted that they may be turned inwards to rest on the bar when not required. It is often necessary to set an object at an angle across the stage during observation, in order that some special feature may appear vertically in the centre of the field. If the springs are not provided this cannot be done.

FINDERS FOR PLAIN STAGES.—The form of finder suggested by the late Mr. Lewis Wright for plain stages is the most efficient for practical purposes. Many proposals have been made, but

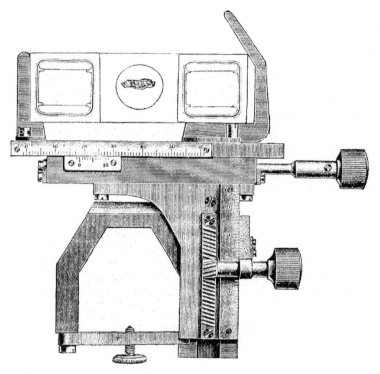

FIG. 4.—ATTACHABLE MECHANICAL STAGE, BY CARL ZEISS.
(Two-thirds full size.)

none equal this one for simplicity. On the right-hand side of the central aperture, one inch of the stage is divided into fifty parts in vertical and horizontal directions. A special feature of interest in an object having been discovered, the slide being maintained in a horizontal position across the stage by means of the sliding-bar, it is only necessary to read from the top right-hand corner of the slide the lines against which it lies. A note of same is made on the labels of the object, and the specimen can subsequently be placed in exactly the same position, and the subject re-examined. Without the sliding-bar it is somewhat difficult to keep the object exactly straight across the stage, but with care, on observing an important feature, the slide can be gently turned until it is in a correct position for taking readings.

The great saving of time that is afforded by such a device as this should establish its claim to be placed on every student's microscope. Several makers have already adopted the arrangement, and it would be a great gain to microscopists in general if a uniform position for the divisions were agreed upon between them, so that a person noting a special point with his microscope could send the specimen, with the readings marked upon it, to a brother worker, and he, having the same kind of finder on his stage, would at once be able to find the desired spot.

The following method would be suitable for the average size of stage : A piece of metal the same size as an ordinary glass slip (3 × 1 inches) should be adopted as a tool, and $\frac{3}{4}$ inch from one end and $\frac{1}{32}$ inch from the edge a minute spot should be made with a small drill. The metal slide should be placed on the stage with the spot towards the front, and the $\frac{3}{4}$-inch space to the right of the centre of the stage. The drilled spot should then be placed central in the field of a 1-inch objective, and the outer margin of the square of divisions marked off from the right-hand end of the metal slide.

The Wright's finder is obviously unsuitable for any other than a stage whose upper surface does not travel.

In selecting a stage for a microscope our choice would therefore be as follows :

For a first-class microscope : Mechanical movements ; concentric rotation ; screws to make the rotation quite true with any objective ; sliding-bar to top plate and stop-pin for object to go against ; divisions to plates of stage reading to parts of millimetre or inch ; rackwork rotation to stage ; and (optional) divisions to periphery of stage.

For a second-class microscope : Mechanical movements ; sliding-bar to top plate ; non-concentric rotation.

For a third-class microscope : Plain stage, with springs to hold object in position ; if provided with sliding-bar or plate as object-carrier, so much the better.

The Sub-Stage or Fitting under the Stage to carry Condenser, etc.

THE SUB-STAGE.—This consists of a tube which should be 1·527 inches = 38·786 millimetres, or, roughly, 1½ inches full, internal diameter — termed the 'Society's size.' It carries illuminating apparatus for condensing the light on the object,

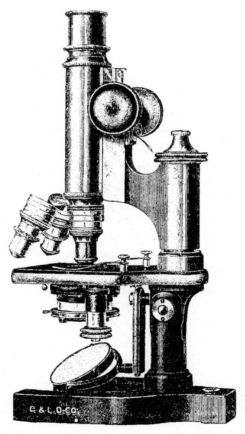

FIG. 5.—BAUSCH AND LOMB RESEARCH MICROSCOPE STAND BB, ON HORSESHOE PATTERN FOOT.

the polarizing prism, and other apparatus, referred to on a later page. It is adjusted in a vertical direction to and from the under surface of the stage by means of a rack and pinion, and the ring carrying the apparatus is mounted in an outer collar provided with screws, by means of which the condenser, or other

2

apparatus, can be made exactly central with the objective with which it is working. This central fitting is made to rotate by rack and pinion in some instances for using the polarizer, etc., but this is so rarely needed that it is unnecessary except for special work. It is essential that the sub-stage should be substantially made, as it is a most important fitting, often too little appreciated. A fine adjustment, to permit of the condenser being focussed in the most exact manner, is often provided with the best stands, and it is exceedingly convenient and of very great importance where high-power work is intended to be done. Often it is wished just to alter the focus of the sub-stage condenser very slightly. In attempting to do so the tension on the milled head of the rackwork is apt to cause vibration, so that the best point of adjustment cannot be at once observed. By communicating this small amount of motion with the fine adjustment the focus is obtained to a nicety. It is also especially convenient where a number of specimens have to be examined. The varied thicknesses of the slips necessitate a slight readjustment of the condenser in each instance, and this can be very quickly done if a fine adjustment be fitted to the sub-stage. Further, the modern sub-stage condensers possess such large apertures that their exact adjustment becomes equal in importance to the precise focussing of the objective.

Where a microscope is provided with a sub-stage it is necessary to ascertain if it will centre with the objective by means of its screws; this should be done in the same manner as described below for the 'under-fitting,' and the centring screws turned. Also, it is very important that when the sub-stage is racked up or down it should maintain its centre with the optical axis. But few instruments will stand this test; in consequence of untrue mounting or building the sub-stage goes out of centre—slightly in some cases, considerably in others. There ought to be absolute truth if everything is square, and any great deviation in this respect should call for rectification. If a fine adjustment be fitted to the sub-stage, it may be tested by using the upper surface as a stage and placing the object on it; this may be made to adhere with a little tallow or grease. An objective of medium power will probably not focus, the sub-stage being too far away. The nosepiece end of the microscope must therefore be lengthened ; for this purpose remove the prism from an analyzer fitting, and use this fitting as a lengthening adapter. The object is then viewed in the usual way.

In the construction of the sub-stage once again the Continental microscopes are not for a moment to be compared with their English contemporaries. It is exceptional for a maker of microscopes on the Continent to provide centring screws to his sub-stage. It is simply impossible to do good work without this convenience. There is hardly a worker nowadays who does not have objectives by more than one maker, and it will be found that these have different centres. Continental opticians maintain that their method simplifies work, for the condenser is centred once for all to the objectives that are supplied with the instruments. Simple and obvious tests will quickly demonstrate that this is quite fallacious.

THE UNDER-FITTING.—In the cheaper instruments, instead of the sub-stage as above described, an ordinary plain tube, termed the under-fitting, is screwed into the under side of the stage, and in this tube the condenser or other apparatus is moved up and down to focus. It is shown fitted to the microscope figured on p. 7. This must be truly centred with the optical tube, and it is well to test it by placing a small diaphragm in the under-fitting, and with an objective in the body to focus the diaphragm. If it is not central it is practically useless. The additional convenience and necessity of the centring sub-stage cannot be too fully impressed upon the beginner who contemplates doing thorough work.

Mention must be made of a useful addition to the under-fitting tube, which consists of a simple but effective means of raising and lowering it so that the condenser may be focussed. The under-fitting is mounted on an arm which is strongly attached to a substantial screw fixed to the under side of the stage, the turning of which effects the purpose. It will be seen on the microscope figured on p. 17. Its efficiency and that of all under-fittings would be greatly increased if centring screws were provided.

The great convenience will be found in many instruments of being able to swing the sub-stage aside out of the optical axis of the instrument on a hinge-joint fitting. It saves much time to students, especially where two or three powers are constantly being interchanged, and the condenser may not be required for all of them. Where this arrangement exists it should be adapted in a workman-like and substantial manner, and a proper support given to the fitting when in the optical axis to make it perfectly rigid.

The choice with regard to a sub-stage would therefore be—

In a first-class microscope: Sub-stage, having rackwork and fine adjustment for focussing, and provided with facilities for centring; rackwork rotation, if for examination of crystals or for petrology.

Second-class instrument: Sub-stage, having rackwork and centring adjustments, and means of lifting aside out of the optical axis.

Student's instrument: The same as the second-class, or with the plain under-fitting, with or without screw focussing arrangement; preferably with centring arrangement. In any case it is imperative that it shall be of the 'Universal' size.

Arrangements for Focussing.

Although other parts have their relative value in producing efficient working, the most important are those which provide the means for focussing the object-glass. Those who have been troubled with movements which have exhausted patience and prevented the execution of work in hand, will sympathize with the insistence on sound principles of construction and perfect workmanship.

Two adjustments are provided for focussing—one called the coarse adjustment, and the other the fine adjustment. The former is usually so well made that it is possible to focus high-power objectives precisely with it. It is by itself sufficient for low-power work. The use of the fine adjustment is for bringing to the sharpest possible focus the details of the objects which are being examined with high-power objectives, and when it is mentioned that the amount of movement required to be imparted is frequently only $\frac{1}{5000}$ of an inch, the exactness necessary will be in some degree appreciated.

THE COARSE ADJUSTMENT.—There is only one type of coarse adjustment now fitted to all instruments, and this consists of a rack and pinion actuating the body in a very true-fitting dove-tailed bearing, as per Fig. 6 shown on p. 21. In the illustration it will be seen that the rack is cut diagonally, and the pinion corresponds. In order that it may work at its best, each tooth of the rack has to be carefully ' ground in '—that is, fitted to a leaf in the pinion—and so that the fitted tooth of the rack may always engage the correct leaf of the pinion, it is necessary so

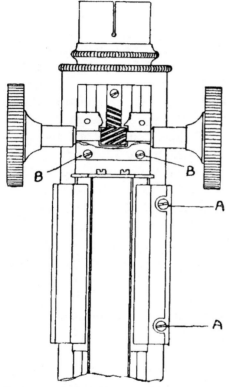

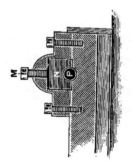

FIG. 6. — SHOWING THE ARRANGEMENT OF RACK AND PINION, AND FINE ADJUSTMENT DOVETAILED FITTING WITH ADJUSTING SCREWS A.

FIG. 7.—SECTIONAL VIEW OF ADJUSTABLE FITTINGS OF RACK AND PINION.

to fix the body that, when racked up as high as possible, it may not be withdrawn from its bearings and rackwork, it being usually provided with a ' stop ' screw. The pinion should have suitable provision by means of adjusting screws for exactly controlling the stiffness of the rackwork action, and for taking up slight backlash which may arise in consequence of wear and tear.

An illustration of the method adopted to secure this result is shown in Fig. 7, p. 21, in which the pinion P is held in position by a block of metal, N, against which pressure is exerted by two screws, one of which, M, is shown in the figure. With this arrangement the most exact relation of pinion to rack can be established and maintained.

For the purposes of the amateur a long range of coarse adjustment is necessary, so that low-power objectives can be used. It should allow of a distance of 3 to 4 inches between the nosepiece and the stage when the body is racked to its highest point. The shortness of the movement in the instruments of Continental manufacture renders these instruments generally unsuitable for the all-round work of the ordinary microscopist.

It has been recommended that microscopists should take their instruments to pieces in order that they may judge of their workmanship; but in reality a well-made microscope requires to be as carefully put together as a watch, and for a novice to attempt to undo the parts means very probable detriment to the instrument. The name of a first-class maker on an instrument may generally be considered a guarantee of good workmanship, otherwise he could not possibly maintain his reputation.

Formerly cheap students' microscopes, instead of being provided with a rack and pinion for the coarse adjustment of the object-glass, were made with the body to slide in a fixed tube. This is a very rough-and-ready arrangement, and accuracy of centring cannot be maintained as with a rack and pinion.

Fine Adjustments.

Well-defined attempts have been made by nearly all makers to improve this most important of all movements. The demand for accuracy in this particular has been greatly increased by the growing use of objectives of large aperture which cannot be profitably employed excepting with a fine adjustment of the utmost precision. Its essentials are that it impart a very slow motion and be absolutely free from lateral movement. The fine adjustment that for many years has proved thoroughly satisfactory in the writer's hands is that made by Messrs. Watson and Sons, also supplied by other makers, in some instances

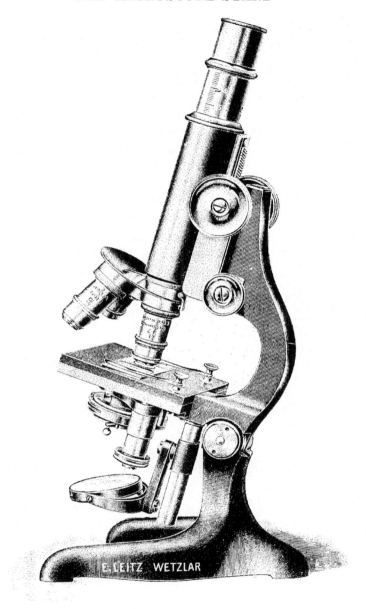

FIG. 8.—STAND F. BY LEITZ, WITH MODIFIED HORSESHOE FOOT.

with modifications. It is shown in position on the instrument (Fig. 1) and the working details will be gleaned from Fig. 9.

The body is raised or lowered in a dovetailed fitting by means of a lever contained within the limb of the instrument, and a pin passing through it transversely acts as a fulcrum. By turning a milled head attached to a micrometer screw, force is applied to the lever at one end against a pointed rod, attached to the body and entwined by a coil spring, at the other extremity. As the body moves upwards, the spring is compressed against a brass plate, and on the micrometer screw being released this spring produces the reactionary power. One arm of the lever is four and a half times longer than the other, consequently the weight of the body at the milled-head end of the lever and the motion imparted are reduced in this ratio. Thus the makers give the weight of a body of one of their instruments as 17 ounces, and this divided by $4\frac{1}{2}$ reduces the resistance to $3\frac{7}{9}$ ounces. It will be seen from this that when in operation, if perchance the objective touch the thin cover-glass of the object, it will do so with a pressure so gentle and slight that there is little risk of damage to either. This system has the advantage that the position of the milled head on the limb is convenient for manipulation, and is not altered when the body is racked up—that is, it is not carried by the rackwork, as in many forms, so that its attachment to a focussing rod of a camera for photo-micrography is easy and convenient. There is also a very simple means of adjustment provided for taking up any slackness through wear. The slide in which the fine adjustment is fitted has slots, to which are fitted screws (shown in Fig. 6, p. 21, marked A). By turning these screws slightly, the spring-fitting grips the bearing-part more tightly, and so takes up any wear caused by friction. Any microscopist can thereby adjust his own instrument.

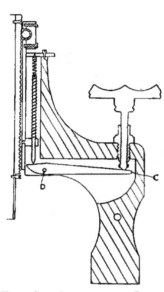

FIG. 9.　SECTION OF LIMB, SHOWING CONSTRUCTION OF LEVER FINE ADJUSTMENT, ETC.

C, lever ; D, fulcrum of lever of fine adjustment.

For simplicity, freedom from complexity, directness of action,

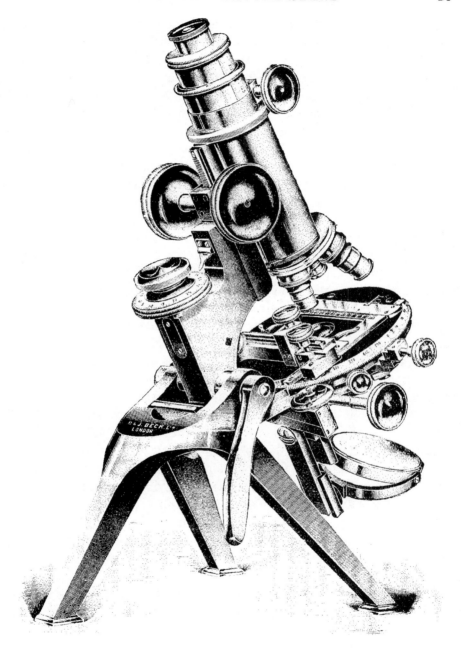

Fig. 10.—'Imperial' Microscope, by R. and J. Beck, showing Two-Speed Fine Adjustment.

and complete effectiveness and convenience, it is unsurpassed. Makers of stands in which a direct acting screw is fitted, as in the Continental form, have exercised considerable ingenuity in providing an efficient movement. Carl Reichert, of Vienna, who has in recent years shown highly progressive tendencies, designed a lever form of fine adjustment for this latter description of microscope, which has proved even more satisfactory than was originally anticipated. Messrs. Swift and Son, too, have adopted a somewhat similar plan with corresponding success in their 'Ariston' pattern, while Messrs. Zeiss and Leitz have devised entirely distinct fine adjustments for their stands, actuated by means of gear-wheels, which produce an extremely slow and precise movement.

Powell and Lealand's instruments are also provided with a fine adjustment having special merit, consisting of a lever actuating a long tube sliding up and down inside the body.

Two-Speed Fine Adjustments.—Complaint is made that the tendency to fit very slow-acting fine adjustments has become a source of inconvenience to students and others who require to work rapidly with objectives of different powers fitted on revolving nosepieces. To meet this Mr. A. Ashe, of the Quekett Club, designed a two-speed fine adjustment, and the plan was carried to a practical issue by Messrs. R. and J. Beck, who now apply it to certain of their instruments. It is shown fitted to the microscope (illustrated p. 25). It will be noticed that two milled heads are provided for fine focussing instead of the usual one. The upper milled head turns a screw having a coarse thread, moving the body $\frac{1}{60}$ inch for each revolution; the lower actuates a fine screw which causes a movement of the body $\frac{1}{300}$ inch for each complete turn. At any moment either milled head may be used and time thereby saved when low powers are employed.

Where a simpler method may be desired, practically the same result can be secured by having attached to the centre of the ordinary milled head a smaller spindle, say, one-sixth of the diameter of the former. When a quick movement is required this spindle can be turned between the fingers, and a rate equal to six to one of that obtainable by slowly rotating the ordinary milled head will be secured. Messrs. Watson and Sons have applied this to their instruments, and the writer has found the working exceedingly satisfactory. It is shown in Fig. 11.

When testing the performance of the fine adjustment, a central cone of light must be used ; if the light be thrown obliquely there will be of necessity an apparent move-ment in the direction from which the light comes. With central illumination there should be no shake or displacement what-ever in the object when it is focussed.

Nearly every maker has his own system or systems of fine adjustment, possessing features more or less desirable, but they are mostly modifications of those mentioned here. Some firms adopt a most excellent form of fine adjustment for a superior class of micro-scope ; while in the students' patterns the

FIG. 11. — SPINDLE MILLED HEAD FOR GIVING TWO SPEEDS TO FINE ADJUSTMENT.

method employed is dissimilar and oftentimes useless for high-class work. It would be far better that efficiency were not sacrificed in such a manner for the small saving in cost involved.

Above all things avoid the form of fine adjustment which carries the *whole* weight of the body of the instrument, or depresses it against a spring, as in some Continental and cheap students' instruments : these are almost worse than no fine adjustment at all, as they invariably soon work loose in the fittings and cause great annoyance.

The Limb.

The design of the limb of the microscope is of special impor-tance, because it carries the body and is intimately associated with the fine adjustment. It should be of substantial shape and strongly made. In our opinion the pattern which is shown in the build of microscopes on pp. 9, 25, etc., is to be preferred before that shown on p. 17, because in the former additional solidity is imparted to the body fittings on account of there being no separate adjustment which has to act and re-act at the back part—in other words, a limb which carries the body at one extremity, and at the other is acted upon by the fine adjustment through a pillar, cannot in the nature of things be so satisfactory as a limb which carries the fine adjustment instead of being supported by it. Still, it cannot be denied that the method of attaching the limb to the pillar adopted by the majority of the Continental makers, and, for the matter of that, those English manufacturers who include this style of instrument amongst their models, is usually a very substantial one.

In many recent models a handle has been embodied in the limb for carrying the microscope—sometimes it forms a part of the foot pillar. This originated in consequence of disabilities associated with the direct acting pillar fine adjustment. With this the limb is drawn up the pillar on the fine adjustment fitting, when the instrument is lifted by the limb ; and when set down it sometimes happened that the fingers would be caught between the back of the stage and the limb, for immediately the part was released it would go back to its proper position by the pressure on the reacting spring. Incidentally a strain was put on the fine adjustment slide in the lifting process. Generally these handles are not of real value, because they are not sufficiently large for more than two or three fingers, and the balance of the instrument is such that it cannot well be carried upright by its means. Such a contrivance is not required where a fine adjustment of lever pattern, or actuated by gear-wheels, is fitted, because none of the mechanism is interfered with when the limb is used, and appropriately so, for lifting the instrument.

The Body-Tube.

In the majority of microscopes of British make, the diameter of the body-tube is considerably larger than is exactly necessary for the draw-tube which carries the eyepiece. In the large microscope-stands of a few years ago the body-tube had of necessity to be large, because the eyepieces were generally of what is known as the capped pattern, and corresponded in size with the build of the instrument, and the large-sized eyepieces are still used in expensive instruments. The student's, or plain type of eyepiece of small diameter, is, however, far more popular, and this requires only a small fitting. The Continental makers usually provide a body-tube only so much larger than the draw-tube as is necessary to accommodate it ; but the large body-tube is of great convenience for photographic work, particularly when no eyepiece is used. It also permits of the unrestricted effect of the special photographic lenses that are sometimes used with a microscope, to be obtained, because when the draw-tube is removed for such purposes there is no 'cutting-off' of the

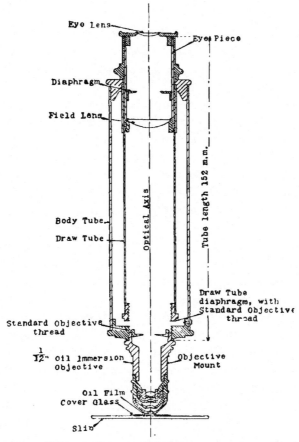

FIG. 12.—SECTIONAL VIEW OF BODY OF MICROSCOPE, SHOWING CONSTRUCTION OF VARIOUS PARTS AND METHOD OF MEASURING BODY-LENGTH APPROXIMATELY.

marginal rays by the body. Further, no worker knows what he may ultimately wish to do; but, having a large diameter of body-tube, he can at any time provide himself with a draw-tube which will carry eyepieces of larger diameter. The large outer body-tube is therefore recommended.

The length of the body has next to be considered. There is a growing tendency to diminish the total length of the body, and with it the extension that is possible with the draw-tube, for obviously a short body entails a short draw-tube. The length of the tube should be such that when the draw-tube is fully extended, the total over-all length shall be 250 millimetres. This permits of the effective use of a very large range of objectives, and gives latitude for adjustment for thickness of cover-glass. This matter is referred to fully on p. 69.

The accompanying diagram gives a sectional view of a body of a microscope, showing a sufficiently accurate method of measuring the body-length.*

In some expensive microscopes of the best type, in order to give the fullest latitude possible, the body is made with an over-all measurement of about 140 millimetres ($5\frac{1}{2}$ inches). It then carries two draw-tubes, by the extension of which, as much as 310 millimetres (12 inches) can be obtained. One of the two draw-tubes is moved by rackwork and pinion, and permits of great exactness of adjustment, both for tube-length and for cover-glass thickness.

This form of body is coming more and more into use, and will be found a very great convenience to the all-round worker. No precise advice can be given without knowing the work intended to be done, but, generally speaking, the short body with the two draw-tubes is much to be preferred to any other.

The draw-tube usually has a scale of divisions engraved upon it to parts of a centimetre or inch. The object of these divisions is to enable a record to be kept of magnifications at different points of extension, or a note to be made of the lengths of tube that give the most perfect corrections for certain objects and objectives.

In all microscopes of medium or high class, the universal screw-thread should be fitted to the lower end of the draw-tube; where there are two draw-tubes it should be supplied to the outer one. The advantages of this adapter are numerous. A low-power objective can be used in it which it is often impossible to focus on many stands, owing to the compactness of the build and shortness of the movement of the coarse adjustment. With the two draw-tubes, if the outer one have this adapter fitted,

* For detailed explanation see footnote, p. 54.

nearly 10 inches of separation can be obtained between the eyepiece and the objective. It is further useful for carrying the apertometer objective and the analyzer, described respectively on pp. 64 and 108; also the Bertrand's lens for examining the 'brushes' of crystals, and for many other purposes.

It has occurred within the experience of the writer that results obtained on a microscope having a large tube could not be reproduced with the same objective on an instrument having a small tube. This was traced to be due to the diaphragm at the bottom of the draw-tube, and it has since been found that in many students' stands the opening of this diaphragm is as small as $\frac{3}{8}$ inch. This is altogether insufficient, and causes restriction to the passage of rays from the objective. It will be well to see that this diaphragm has an opening of at least $\frac{3}{4}$ inch.

Tailpiece and Mirrors.

The Tailpiece.—It will usually be found convenient if the arm, or tailpiece, which carries the mirrors, be so mounted as to be turned aside with the mirrors when desired. This arrangement is of great utility, for it permits of light being readily directed from the lamp through the sub-stage condenser for critical work. Opticians favour a rectangular rather than cylindrical tailpiece to carry the mirror gimbal; the reason for this is a little doubtful, but there is probably no distinct advantage in one over the other. Where, however, a cylindrical tailpiece is provided, it will be obvious that the mirror could quickly be swung round out of the axis of the microscope and so obviate the necessity for the swinging of the tailpiece itself, but this is quite a minor consideration.

The Mirrors should be plane and concave, hung in a gimbal, giving universal movements, and have a means of adjustment to focus in a vertical direction. The plane mirror is always used with the condenser, spot lens, etc., and with very low-power objectives, but the concave, when the condenser is not employed and the maximum amount of light is desired.

A little experiment will show why this is so. Set up a microscope with a 1-inch objective of moderate aperture, transmit light from a plane mirror, then remove the eyepiece, look down the body-tube, and observe the extent to which the back lens of the objective is flooded with light. Repeat the experiment with a concave mirror. By this means it will be seen that, whereas with the plane mirror the lens is incompletely illuminated, with the concave it is filled with light. It is important that the illuminant shall be in such a position relatively to the concave mirror that when the latter is in use it will be as nearly as possible at right angles with the optical axis of the instrument.

A source of trouble and annoyance in accurate work frequently arises from the imperfect surface of the plane mirror. It sometimes happens that it produces several reflections of the edge of the lamp-flame. This can be to a large extent overcome by the use of a rather more costly parallel-worked mirror, but with this there must be two reflections—the faint image from the upper surface, and the bright one from the lower silvered surface. In consequence of this, critical workers eliminate the mirror, and set the microscope so that the light from the illuminant passes directly into the condenser.

The parallelism of a mirror may be tested by holding it just below the level of the eye in the direction of a row of objects, such, for instance, as chimney-pots; and on observing the reflections, each subject should stand out singly and clearly. If the mirror is not parallel-worked, several reflections of the same object will appear superimposed in the mirror.

Care is needful in the use of the concave mirror, if the best result is to be obtained with it. It should be so arranged that the apex of the cone of rays that it transmits may be exactly in focus on the object. Many microscopes are provided with mirrors that are unsuited to the instrument, being either too long or too short in focus, and consequently do not produce good effects. To test the mirror, a piece of white paper should be placed upon the stage of the instrument, which must be set horizontally, and light from a lamp reflected by the concave mirror on this; then, by sliding the mirror up and down on the tailpiece, it can quickly be seen if the focal point can be obtained upon the paper.

Binocular Microscopes.

We have hitherto been treating principally of the monocular microscope, and this, it must be understood, is the only form that can be used for critical high-power work. Continental firms as a rule do not make binocular microscopes of the usual kind, but two or three of them make a special pattern for dissecting, and a binocular eyepiece, which will be found described under the head of eyepieces. The advantage of a binocular microscope is, that both eyes can be employed simultaneously, saving the strain on the vision which is apt to ensue through the constant employment of the monocular microscope, and the endeavour to see in the best manner the detail in the specimens examined. We should recommend every user of the monocular microscope to train himself to work with either eye, keeping the one not in use open ; this will be found of the very greatest service. The universally understood binocular microscope is provided with a prism, designed by Wenham, which admits of the light going up a direct tube, and reflects light also into a second tube. By this means objects can be seen more naturally than with the monocular microscope, for the reason that stereoscopic vision is obtained, and objects having a certain amount of depth may be seen completely with the binocular microscope, whereas with the monocular it would be necessary to focus in successive stages through the several planes. Especially is this true regarding opaque objects, with low powers. The stereoscopic binocular conveys an impression of the objects viewed that is almost startling in its beautiful effect. Subjects stand out in relief, exhibiting their natural contour, and at once the worker is able to decide the shape and form of an object in a way that it is impossible to do by focussing each separate plane with a monocular instrument. The binocular microscope is *par excellence* the instrument for the amateur. To him the beautiful appeals in a manner that it perforce cannot do to the scientific man, who, being intent on the pursuit of knowledge of some obscure point, has no time to notice, or if to notice, cannot linger to reflect upon the æsthetic aspect. In the examination of rotifers and other inhabitants of 'ponds and rock pools,' perhaps the most charming subjects that the microscope has

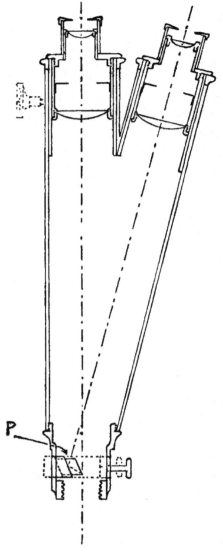

Fig. 13.—Sectional View of Body of
Wenham Binocular Microscope,
showing the Method of sliding
the Prism P out of the Field so
that the Monocular Tube only may
be Used.

ever revealed, the microscopist with a monocular instrument cannot possibly appreciate and interpret structure and movement in the same accurate manner that the binocular enables him to do. These facts should receive careful consideration when a microscope is to be chosen, but it must be borne in mind that the Wenham stereoscopic form of binocular cannot advantageously be used with an objective having a higher numerical aperture than 0·26. Dr. Carpenter some time since pointed out that when an objective having a larger aperture was employed with the Wenham binocular, spherical objects became distorted, and instead of appearing round in shape they became conical. Provision is, therefore, always made whereby the prism may be withdrawn for the higher power objective, and the light then only goes up the monocular or straight tube, and the instrument is to all intents and purposes as useful and convenient as the monocular microscope, while the unemployed eye of the observer is rested by looking into the blank binocular tube; the fact of its not being illuminated will scarcely be noticeable.

For use with the binocular microscope, the closer the

posterior lens of the objective is brought to the prism the better.

It must be understood that all vision through the microscope in the ordinary way is inverted, that is, the object is seen upside down. A very good form of binocular microscope, devised by Stephenson and made by Swift and Son, erects the image, or, in other words, enables it to be seen the right way up. It is excellent for dissecting purposes, and high powers can be employed with it; still, it cannot be described as an all-round microscope, and would have to be classed with instruments for special work. Our advice on the question of a monocular or binocular microscope is: If the instrument be required for strictly educational, scientific, or photographic work, the monocular must be chosen. The bulk of the general amateur's work is done with comparatively low powers, and in such cases the binocular is unquestionably of advantage, and to be preferred. If it is proposed to combine scientific with general work a good plan is to have two separate bodies—monocular and binocular—interchangeable in the same bearings. The maximum facility is then at the disposal of the user. It should be noted that when the two bodies are chosen, it is well to have centring screws to the rotating stage, as described on p. 11, because the bodies rarely have identically the same centres, and the stage could not otherwise be made to rotate concentrically with both bodies.

Mention is made on a later page of the Greenough binocular microscope. It is not an instrument that is suited for general work, being primarily designed for the preparation of objects.

Microscopes for Special Purposes.

DISSECTING MICROSCOPES.—For serious work, very elaborate instruments have been devised, and are to be found in the biological laboratories. A well-known instrument of this class made by Carl Zeiss, and completely fitted with its auxiliary apparatus, costs over £20. Much good work is, however, performed with comparatively inexpensive instruments, but it is important that the rests for the hands are sufficiently long to give the necessary steadiness. Support must be given to the forearm as well as the wrist, and when the elbows are on the table, the forearm and the wrist should be carried upon the arm-rests of the instrument. Although this may appear to be a small matter, it is really of great value, both from the point of view of comfort and of good working. Next it is desirable that the stage shall have a sufficiently large surface to accommodate the dish in which the subject for dissection is placed. The stage may have a surface of glass or ebonite. The arm carrying the magnifying lens should be freely movable and preferably double-

FIG. 14.—A SIMPLE PATTERN OF DISSECTING MICROSCOPE.

jointed ; rackwork and pinion to focus the lens is also a necessity. The light-reflector should be carried in a gimbal, and consist of a mirror on one side and a mat opal disc on the other.

A popular form of dissecting microscope, made in large and small sizes by many opticians, is that shown in the accompanying illustration (Fig. 14) ; and although it has several of the defects referred to, experience has shown that the general design is a useful one.

With the aplanatic magnifiers that are usually employed with the dissecting microscope the image is seen erect—that is, the right way up—but the magnifying powers are necessarily limited to those which give a sufficiently free space for the use of the dissecting needles, etc. The use is therefore restricted to comparatively low powers. If the compound microscope is used— that is, an instrument with an eyepiece as well as an objective —the image is seen inverted; but greater working distance, with higher magnifying powers, is secured by this combination. The inversion of the image, however, rendered its use impracticable. The introduction of the Porro prism erector changed this. This consists of a body fitted with right-angle prisms in the same manner as the modern prism binocular glass, carrying an eyepiece at one end and an objective at the other, the prisms serving to present the image the right way up, and to permit of the reduction of length of mounting between eyepiece and objective. Increased magnifications are obtained with this arrangement, with ample working distance, and it has proved of invaluable assistance. It is shown on the excellent type of microscope by Zeiss on p. 38.

Much useful work can be done by means of a simple upright stand having a horizontal arm, preferably with a simple joint, and always with provision for raising and lowering the arm. The magnifier is carried in a suitable fitting on the arm, and frequently fills all the requirements of the amateur worker.

MAGNIFIERS FOR DISSECTING.—These are usually made on a plan originated by Steinheil, consisting of three lenses cemented together, and the perfection to which they have been brought is remarkable. There is no occasion to indicate special makers of these lenses; all opticians of repute supply them in good quality. It is important to remark that magnifiers consisting of single uncorrected lenses, in which no attempt at aplanatism or achromatism is attempted, are not to be recommended for other than very elementary work. Only the centre of the field of such a lens is ever useful, while with the aplanatic form the subject is sharp and clear to the edge. The various magnifiers are also supplied in mounts for carrying in the pocket, and will be found useful aids to the microscopist.

There is a frequent request from the unknowing for magnifiers having large lenses and yielding high magnification. Such a combination is impossible. It will be understood when it is

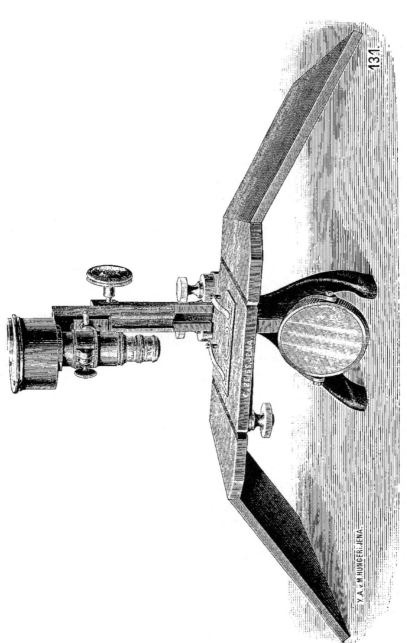

131

Y.A.v.M.HUNGER:JENA.

FIG. 15.—MONOCULAR PRISM DISSECTING MICROSCOPE BY ZEISS. (One-third full siz)

recalled that the magnifying power of a lens depends upon its focus, and the focus, again, upon its radius of curvature. For this reason, a lens of 1 inch diameter having a focus of 1 inch would be a hemisphere; and as the distortion with an uncorrected hemispherical lens would be enormous, and only a small part of the centre of the field reasonably sharp, it will be realized that with this maximum curvature of a simple lens no useful effect would be obtained. It is therefore necessary that the lens should be of smaller diameter than twice its radius. For instance, a lens of ½ inch diameter and 1 inch focus would be more practical in working. To merely call attention to this matter will be sufficient to show that a large diameter and high magnification cannot be associated with a simple lens.

FIG. 16.—THE APLA-NATIC MAGNIFIER.

Special reference must be made to the Greenough binocular microscope, which is a speciality of the firm of Carl Zeiss. It is not an instrument that is suited for general work, being principally designed for the preparation of objects. The Porro prism-erecting system is employed for the body, which consists of a combination of two incorporated bodies side by side, and two special objectives accurately paired and mounted at such an angle to each other that they both embrace exactly the same portion of the field. This, it will be noticed, is quite different from the ordinary Wenham binocular microscope, in which a single objective is used, and the division of the light into the two bodies is effected by means of a prism.

Only a limited range of objectives can be applied to the Greenough instrument; these in each case have to be specially arranged and supplied on an appropriate carrier, but they can be had in a range of magnifications from 8 to 72 diameters.

As before remarked, the Porro prism system enables the object to be seen the right way up; but, beyond this, a beautiful stereoscopic effect is obtained—superior, probably, from all points of view, to any other different arrangement.

Microscopists who have been accustomed to the binocular microscope, viewing a natural history specimen through the Greenough binocular, have never failed to express astonishment and pleasure.

Microscopes for Metallurgical Work.

Examination of metals by means of the microscope is a comparatively modern study, but there is probably no iron or steel works of standing that is not equipped with suitable instruments both for observing and photographing. By means of the micro-

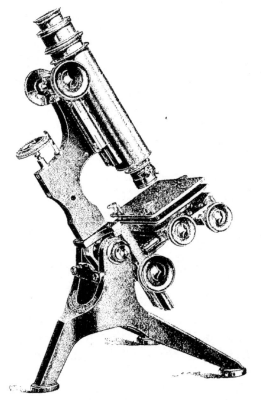

FIG. 17.—SIMPLE PATTERN OF METALLURGICAL MICROSCOPE, SHOWING ADJUSTABLE STAGE.

scope much information regarding both the chemical constitution and mechanical properties are disclosed, but it is especially valuable for the latter. For instance, the structure of steel varies with the degrees of hardness and the amount of heat to which it has been subjected, and it is possible readily to gain definite information concerning the suitability of the metal for the service to which it is to be put by means of the microscope. In the manufacture of guns any defect which may have taken place in the heating or quenching of the steel, which would render the gun unsafe or unsatisfactory, can be discovered before the manufacture is proceeded with. Engineers can detect flaws, blow-holes, defective welds, etc., at an early stage, and avoid the trouble incident to the use of imperfect metal in the finished article.

The main support of the stage is carried in a dovetailed fitting, parallel with the body of the instrument, and can be raised or lowered by means of a rackwork and pinion.

The stage itself has on the upper surface a levelling plate, on which the specimen for examination is laid. Three screws permit of any want of parallelism between the faces of the specimen being compensated for.

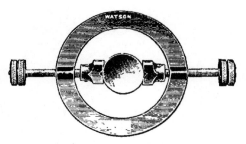

To illuminate the specimen, a vertical illuminator, the construction and use of which is referred to on p. 113, and which is fitted between the objective and the nosepiece, is employed for

FIG. 18.—HOLDER FOR METALLURGICAL WORK.

medium and high-power examinations. The lamp and bull's-eye have to be placed in fixed relation to this vertical illuminator, and it is important that, once the illuminant is adjusted, no further movement should take place; this renders obvious the utility of the rackwork for raising and lowering the whole of the mechanical stage, and the addition of a fine adjustment to this part of the stage becomes an added convenience.

The use of this class of microscope is by no means restricted to iron and steel examination. Similar instruments are largely used for the examination of brass, in addition to other metals, and a large quantity of alloys and compositions, such, for instance, as certain material of which billiard-balls are made.

It will be obvious that the subjects examined are opaque, and the usual mirror and sub-stage apparatus are dispensed with. To conduct such work in the most advantageous manner, the microscope should be specially designed for the purpose and kept to it. Long experience produces the conviction that special work of such a nature should be done with an instrument devised for the particular purpose, rather than by attempting to make an ordinary instrument serviceable. So extensive has this work become, that all leading makers construct metallurgical micro-

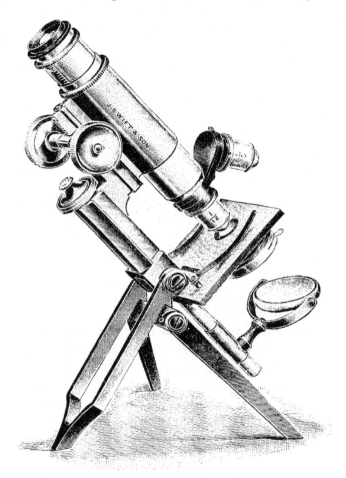

FIG. 19.—TRAVELLER'S MICROSCOPE.
By J. Swift and Son.

scopes. The patterns vary, but the main features will be gleaned from the illustration of one of these stands on p. 40.

For some purposes a holder for gripping the metal under examination has been found advantageous, and this is particularly the case if a microscope of the ordinary type is available. There are several different types, but the one depicted in Fig. 18

is a simple one, which can be made to carry a fairly large piece of metal between its adjustable jaws.

If an ordinary microscope is employed for metallurgy, it will be found convenient to use the sub-stage as the object-carrier instead of the stage.

The vertical illuminator will generally give the requisite length to bring the objective into focus if the specimen be carried on the sub-stage, and the latter racked up nearly as high as it will go. It will be seen by this arrangement that the light can be adjusted to the vertical illuminator, and the focussing done by means of the sub-stage; the metal-holder previously referred to, or a special plate with a fitting to go in the sub-stage, will thus enable occasional work to be done.

For low-power work with the 1-inch objective or less magnification, the ordinary bull's-eye or stand condenser, together with a parabolic reflector (referred to on pp. 110 and 113), are used. Sometimes the latter is used in conjunction with an additional reflecting plate, arranged in a manner devised by Sorby.

It requires to be borne in mind that records have frequently to be made photographically of metallurgical specimens after visual examination, and due consideration must be given, in the selection of an instrument, to its suitability for the combined purposes; particularly, when it is set horizontally, there must be no tendency to unsteadiness; it must be equally firm in all working positions.

Portable or Traveller's Microscope.

A lover of the microscope feels lost on his holidays, or when travelling, without his microscope, and numerous attempts have been made to produce instruments that shall have the satisfactory working qualities, and yet, by folding and detaching parts, occupy a minimum of space. The necessity of such instruments has been appreciated by the numerous investigators of tropical diseases, and this type of microscope has played no inconsiderable part in this class of work and in clinical and medical observations. From the amateur's point of view, one of the most successful instruments is that by Swift and Son, shown in Fig. 19, which is fitted into a case, with objectives, eyepiece, and several useful accessories, measuring $9 \times 3\frac{1}{4} \times 3$ inches. Excellent models are also made by C. Baker, R. and J. Beck, Leitz, and W. Watson and Sons, the last two named also offering instruments of heavier type, rendered especially compact for high-power work.

Petrological Microscopes.

The examination of rocks and minerals has become a department of such great prominence that a larger number of special microscopes have been introduced for this specific purpose during the last decade. Swift and Son, who are specialists in this particular class of instrument, offer no less than seven different models and instruments of excellent design, all having distinctive features, and others are manufactured by R. and J. Beck, Leitz, W. Watson and Sons, and Zeiss.

With such a variety, it is difficult to indicate the essentials without knowing the precise work that is to be undertaken. Very simple arrangements will suffice for occasional work, while every elaboration possible is necessary for the more gregarious worker. Perhaps it will be well to mention the way in which the ordinary microscope may be employed in petrology in order to indicate what is necessary. First, the stage should be a concentric rotating one, with the circumference divided to degrees, reading by a vernier or pointer, so that the crystal under examination may be measured and its angle ascertained. The polarizer, having a large-size prism, must be used beneath the stage, and the rotating flange should be divided and have an indicator for reading.

For a certain class of work, a condenser of large numerical aperture would require to be adaptable to the polarizer. The analyzer can be fitted above the eyepiece also with a divided circle. The quartz wedge can advantageously be placed in a carrier between the objective and the nosepiece, and the Bertrand's lens for observing the interference of figures of crystals screwed to the lower end of the draw-tube. A special eyepiece would have to be provided having cross-webs fitted to the diaphragm, and slots should be made either in the eyepiece or immediately above it to receive the undulation plate, micrometer, etc.

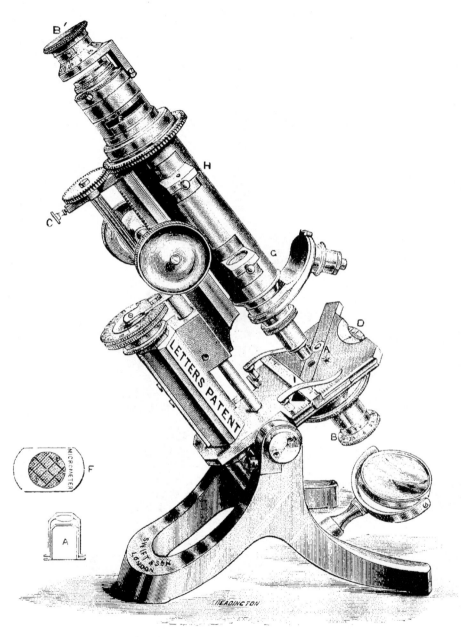

FIG. 20.—PETROLOGICAL MICROSCOPE.

By Swift and Son ; designed by Mr. Allan Dick.

Now, all these fittings can be applied to an ordinary microscope at any time, but where petrological study is regularly undertaken, it is far better to be equipped with an instrument built for the

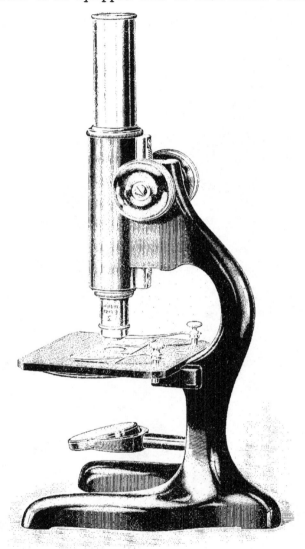

FIG. 21.—STAND VI., BY LEITZ.

A very inexpensive microscope, suitable for preparing and mounting specimens.

specific purpose, with all these conveniences incorporated in a manner that enables full control and rapid manipulation to be exercised.

The microscope illustrated and designed by Mr. Allan Dick has been a standard model for many years. The striking feature in it is that the stage is fixed, and instead of rotating the specimens, the polarizer and analyzer prisms with the eyepiece are made to revolve together. It is not universally agreed that this arrangement is the best, but it has been much favoured, and it will be at once seen that the trouble connected with the exact centring of the rotating stage is obviated by this means.

The microscope in petrology is dealt with in a separate chapter (p. 219), and reference should be made to this for further information.

Microscopes for Preparing and Mounting.

A plain substantial microscope-stand, which the user would have no compunction in soiling, is of inestimable value in the preparing and mounting of micro-slides, and a suitable one can be obtained for such a small sum that it is a pity that instruments designed for work of a higher grade should be employed for the purpose. We figure on p. 46 an illustration of the type of microscope that will be found very serviceable for this purpose. When we mention that the cost of this is only 20s., that it is provided with rackwork and pinion, and that it has the standard screw for objectives, and receives eyepieces of Continental diameter, it will probably need no further recommendation.

The Ultra-Microscope.

This interesting instrument was invented by Dr. H. Siedentopf, a member of the scientific staff of the Jena Glass Works, and is manufactured by Carl Zeiss. It renders visible particles which are beyond the limits of vision with the ordinary microscope, even with the highest powers, and are hence described as ultra-microscopical.

It has, of course, long been known that whilst there is a fairly-well-defined limit of resolving power—in other words, a minimum distance beyond which two or more particles or elements of structure cannot be seen apart—there is no such limit for the size of the separate particles themselves, at any rate when they are placed in a strong light against a dark background; for under these conditions the smallness of the reflecting surface of the particles can be compensated for by the brightness of the illumination in much the same way as minute particles of floating dust become visible in the path of a ray of sunlight through an otherwise shaded room.

The effect is obtained in a most ingenious manner by taking advantage of total reflection. The condenser (a dry one) is arranged at right angles to the optical axis of the microscope, so that its concentrated light enters the slide containing or forming the object laterally—through the edge. None of this light can emerge into air through the upper surface of the slide, as this would demand a refractive index of 1·414—lower even than that of fluorite. The light diffused by small particles is, therefore, the only light that reaches the eye, and the latter will therefore be sensitive to its utmost limit.

But it is further necessary to limit the illumination to particles within the depth of focus, as otherwise the light scattered by particles not in focus would form a luminous background. This object is attained by focussing the source of light—generally the electric arc—on an adjustable spectroscope-slit, then focussing the slit in the object.

By the use of this refined method Dr. Siedentopf proved by direct observation that the beautiful red glass, known as gold-ruby glass, contains small, detached particles of gold of extremely small size. By computing the cubic space under observation, counting the particles within that space, and ascertaining by analysis the percentage of gold in the glass under examination, it was found that the particles were equal in weight to tiny cubes of gold having sides equal to from one-sixth to five one-millionths of an inch. Though almost inconceivably minute, the particles are still composed of a great number (at least 10,000) of molecules. These latter must, therefore, still be considered as invisible.

By its means very minute particles of other origin can also be demonstrated—such, for instance, as the flagellæ of certain bacteria.

Reference is made to this subject, not by reason of its importance to the elementary worker, but because it shows that the possibilities of the microscope have not yet reached their limit ; and, although at present, work with this instrument is so specialized that it does not come within the range of the ordinary worker, future developments may take place on similar lines which may make it more generally useful.

Modified effects of the nature indicated are obtained by the use of a recently introduced illuminator, known as the 'immersion paraboloid,' the 'paraboloidal condenser,' etc. These are fully described on p. 105. They are sometimes inaccurately described as 'ultra-microscopes,' but the construction is quite dissimilar, and the titles should not be indifferently applied.

CHAPTER II

OPTICAL CONSTRUCTION

Preliminary Note.

IN the former part of this book we have dealt exclusively with the stand, or mechanical means of employing the optical system and accessories ; and important as it is that those details shall be very efficient, it is, if anything, still more so that the eye-pieces, objectives, and illuminating apparatus shall be of the most perfect description, properly adapted and intelligently employed, for on the optical combinations depend the results that are to be obtained with the stand ; and although care and trouble may enable a person to use a bad stand, no good stand can ever compensate in any way for bad objectives. It requires constant practice and a long apprenticeship to learn to use the microscope to the utmost advantage. Every subject of examination calls for special manipulative treatment if it is to be correctly understood and appreciated. Experience alone can guide in obtaining the best result under varied circumstances, and that experience must be based on a knowledge and understanding of correct methods in working.

Definitions.

Some of the following terms will be made use of in this book, and are constantly met with in literature on microscopical subjects ; a brief explanation of them may therefore prove of service.

Achromatic Correction.—Owing to the relatively greater dispersive power of flint-glass (containing lead or other heavy metals) as compared with crown-glass, it is possible to produce a combination of a convex lens of crown-glass with a concave lens of flint, which collects rays like a simple convex lens, but which unites two different colours in the *same* focal point, thus in a great measure correcting the chromatic aberration.

Aplanatism.—A freedom from spherical aberration (see below).

Apochromatic Correction.—The highest attainable correction of microscope objectives, comprising the correction of spherical aberration for all colours, and the union of *three* different colours in one focus, thereby eliminating the secondary spectrum.

Chromatic Aberration.—White light is the composite effect of a continuous range of colours, passing from red, through yellow, green, and blue to violet (see Spectrum). All transparent media have different refractive indices for these different colours, and, as a consequence, after their passage through a simple lens the rays do not unite at one focal point. The red rays, being the least refrangible and bending to the smallest extent, unite at the farthest distance from the lens; the orange and green rays unite at points closer to the lens; while the violet rays come to a focus at a point nearest to the lens. The confusion of different coloured images resulting from this dispersion is termed 'chromatic aberration.'

Chromatic Over-Correction.—A term used when a lens brings yellow or even orange rays to the shortest focus and best correction.

Chromatic Under-Correction.—A term applied to a lens when rays towards the blue end of the spectrum are best corrected. Thus a photographic lens is *visually* under-corrected.

Diaphragm.—This is generally understood in optical instruments to be a circular opening in a plate that is used to cut off the marginal portions of a beam of light, and in this sense is referred to in this book. The diaphragm is often improperly called a *stop*.

Diffraction Spectra.—If we look through a finely ruled grating at a gas or candle flame, we shall see a large number of images of that flame having the colours of the spectrum. This effect is due to diffraction. In the microscope, objects having fine and regularly spaced markings diffract the light in a similar manner, the resulting diffraction spectra being plainly visible at the back of the objective. According to the Abbe theory of microscopic vision, these diffraction spectra determine the character of the image seen, the latter becoming less like the real structure when the number of diffraction spectra admitted by the objective is reduced, a faithful representation of the object being obtainable only when all diffracted light of sensible brightness is admitted. A further note on this interesting subject by the late Dr. G. Johnstone Stoney, F.R.S., will be found on p. 134.

Female and Male Screws.—The former is a threaded fitting which receives a screw, and the latter a screw which goes into the female fitting. In the case of a bolt and nut the former would have the male, and the latter the female screws.

N.A., abbreviation for ' numerical aperture.' See p. 62.

Negative Eyepiece.—This is an eyepiece for examining an image formed at the diaphragm set between the two component plano-convex lenses. The Huyghenian is the best-known form of negative eyepiece.

O.I., abbreviation for 'optical index.' See p. 68.

Positive Eyepiece.—This is an eyepiece for examining an image situated beyond the field lens. It can consequently be used as a magnifying-glass, etc.

Refractive Index.—When a ray of light passes obliquely from one medium into another of different density, the path of that ray is bent or altered in its course. According to the law of refraction, there is a constant ratio for any given two media between the sine of the angle of incidence (being the angle included between the incident ray in the first medium and the perpendicular) and the sine of the angle of refraction, or of the angle included between the ray after refraction and the same perpendicular. The numerical value of this ratio for a ray passing from air into a medium is called the refractive index of the medium.

Secondary Spectrum.—In an achromatic lens the chromatic aberration is corrected for the brightest (yellow or green) rays of the spectrum, and the pronounced colour shown by uncorrected lenses is in consequence removed. A stricter examination, however, shows that rays of a different colour are not brought to the same focus, for owing to the fact that flint-glass, as compared with crown-glass, disperses the more refrangible rays relatively too much, and the least refrangible relatively too little, a peculiar *secondary* spectrum results from the achromatic combination, the rays corresponding to the brightest apple-green part of the ordinary spectrum, being very closely united and focussed nearest the combination, whilst the other colours focus at increasing distances *in pairs*, yellow being united with dark green, orange with blue, red with indigo. The composite effect of these colours is best seen with oblique light, causing dark objects to have apple-green borders on one side and purple ones on the other.

Semi-Apochromatic Correction.—In achromatic microscope objectives of the older type, chromatic defects that are worse than the secondary spectrum are caused by spherical aberration of the coloured rays, the spherical aberration being corrected for the brightest part of the spectrum only. Objectives made entirely of glass, and therefore showing the secondary spectrum, are called semi-apochromatic when the spherical aberration is corrected practically for all colours.

Spectrum.—A band of colours produced by the splitting up of white light by means of a prism. The order of the colours is: Red, orange, yellow, green, blue, indigo, violet.

Spherical Aberration.—Rays of light passing through the marginal portion of a lens come to a focus nearer to the lens itself than those rays which pass through the centre of the lens, and the interval between the focal points of rays which pass through the marginal and the central parts of that lens is the spherical aberration. In compound lenses this spherical aberration can be corrected for one or more special rays, and a lens so corrected is called aplanatic. It is only truly aplanatic for the particular rays for which it has been accurately corrected.

Spherical Over-Correction is present when a lens unites the marginal rays at a greater distance than the central rays. Spherical under-correction is indicated when the marginal rays focus closer to the lens than the central ones.

Spherical Zones.—In objectives of considerable aperture the intermediate rays may show decided spherical aberration, although the central and marginal rays are united. This defect is meant when spherical zones are spoken of. The degree to which spherical zones are corrected determines chiefly how large a cone of illumination, and how deep an eyepiece an objective will bear before 'breaking down.' A high degree of correction for spherical aberration and spherical zones must accompany the reduction of chromatic defects before terms such as 'semi-apochromatic,' and especially 'apochromatic,' can be applied to a lens.

Stop.—In an optical instrument this is a means of obstructing the passage of the central portion of a beam of light.

MAGNIFYING POWER.

It has been previously stated that magnifying power is not dependent on the size of the instrument. Given suitable eye-pieces and objectives, the same magnification may be obtained on a small microscope as on a large one. It is entirely produced by the two optical parts—the objective, and the eyepiece, or ocular. Under the head of 'Objectives' in the English makers' catalogues it will be noticed that the powers are expressed as 1-inch, ¼-inch, ⅛-inch, etc. The figures do not indicate the distance at which the lenses focus on the object, but are intended to approximately convey the equivalent focus, and thereby the actual magnifying power of the objective. To understand this description, imagine an objective to be placed so as to form an image of an object at 10 inches from its back lens.* Then, an objective which, when so placed, formed an image on the screen which was ten times (diameters) the real size of the object would be described as a 1-inch objective ; one which formed an image twenty times the size of the object would be called a ½-inch objective, and in general the result given, when the magnification of the image formed on the screen is divided into ten, is what is spoken of as the focal length of the objective; also the equivalent focus of an objective divided into ten gives its magnifying power—thus a 2-inch should magnify 5, a ¼-inch 40, and a ⅛-inch 80, diameters. This is termed the initial magnifying power of an objective.

* The above plan will be sufficiently accurate for experimental purposes, but, strictly speaking, it is the equivalent focus of the objective which determines its magnifying power, and in order to obtain exact results the measurements should be taken from the upper focal plane of the objective. The optical tube-length should also be reckoned in like manner, and this may generally be assumed to be from ½ inch to 1 inch longer than the mechanical tube-length or the length of the body of the microscope.

The foci of German objectives are usually expressed in milli-metres, 250 millimetres (about $9\frac{13}{16}$ inches) being taken as the normal vision distance, and the focal length of the objective divided into 250 gives the initial magnifying power.* Thus a 3-millimetre objective should have an initial power of $83\frac{1}{3}$, and a 4-millimetre of $62\frac{1}{2}$, diameters, and so on.

The image formed by the objective is again magnified by the eyepiece. Unfortunately, the latter is rarely marked with its magnifying power, the general rule being to call the different powers by the letters A, B, C, D, etc., or 1, 2, 3, 4, etc. This is not very intelligible, and it would be far better either to express their equivalent focus, as in the case of objectives, or to have the magnifying power in diameters marked on the cap. We will take it that the 'A' eyepiece yields a magnification of 5 diameters. When this, therefore, is used in conjunction with the 1-inch objective on a 10-inch tube-length, which according to the rules previously given would produce a magnification of 10 diameters, the resultant combined power is 50—that is, the powers of the objective and eyepiece multiplied together. The method of estimating the power with short tube-lengths is referred to on p. 72.

OBJECTIVES.

For our purpose we shall divide the subject of objectives into two classes—(1) the apochromatic, and (2) the achromatic. The immersion objective which may belong to either of these two classes is referred to separately. As in our remarks on objectives we shall constantly use the two terms, we will describe their reference.

APOCHROMATIC OBJECTIVES.—The introduction of these objectives by the firm of Carl Zeiss, of Jena, Germany, about twenty-five years ago, placed the science of microscopic optics on a far higher level than had hitherto been attained, and as a result many of the traditional modes of working were altered. Greater precision has been necessitated in the microscope-stand, and the provision of sub-stage condensers of corresponding optical quality to the objectives has been essential. Professor Abbe and Dr. Schott were granted a subsidy by the Prussian

* See note on p. 54.

Government with a view to the promotion of optical research, and aided by this they were able to produce several varieties of new optical glass. The employment of these new glasses in conjunction with fluorite, based upon the careful and elaborate calculations of Professor Abbe, resulted in the production of apochromatic objectives. In these lenses, aberrations which were inherent in the older systems were eliminated or minimized —that is to say, the secondary spectrum was practically removed, and spherical aberration was very perfectly corrected for all colours. The objective, therefore, produced, to all intents and purposes, a colourless image. Higher apertures were obtainable, and in consequence of the improved corrections, accompanied by greater brilliance of the field, the use of eyepieces of high power was rendered permissible and advantageous.

The new kinds of glass were placed at the disposal of opticians throughout the world, and apochromatic objectives have been since manufactured by other firms, whose productions compare favourably with the best of the originators' lenses. The apochromatic objectives by Zeiss have their equivalent focus engraved in millimetres, and it is becoming usual for the same method to be applied to other objectives also. The initial magnifying power of such lenses is ascertained by dividing the equivalent focus in millimetres into 250. Thus, a lens with an equivalent focus of 2·5 millimetres would have an initial magnifying power of 100 diameters.

Special eyepieces, termed 'compensating oculars,' are necessary when using the apochromatic objectives. They will be found described on p. 84.

ACHROMATIC OBJECTIVES.—All objectives that are not actually comprised in the apochromatic category—that is, in which the secondary spectrum is not eliminated—are included under this heading. So far as the principal opticians are concerned, it comprehends a better class of objectives than it did at the period when apochromatic lenses were introduced. By the use of the new optical glasses previously referred to, and in consequence of keen competition amongst manufacturers, many achromatic objectives, tending towards apochromatism, have been made. Several of these are so well corrected that in some instances they vie with the apochromatics in performance.

The class has consequently arisen which has been referred to under the generic term of *semi-apochromatic*. Many of these lenses are made in such perfection as to be superior even in some features to the real apochromatics. Some of the lenses in Watson and Sons' new series of 'Holoscopic' objectives, which require to be used with over-corrected eyepieces of the compensating type, are especially free from spherical aberration. Messrs. Swift and Son, of London, in their series of pan-aplanatic objectives, produce beautiful results; also C. Reichert, of Vienna, and Leitz, of Wetzlar, make objectives that are worthy of special consideration. Beyond these there are excellent series of lenses made by all the microscope manufacturers which meet the requirements of the ordinary amateur in a most efficient manner —in fact, the general quality of such objectives is superior to that which obtained in the so-called best lenses of a few years ago.

ACHROMATIC *versus* APOCHROMATIC OBJECTIVES.—In view of the foregoing facts, it will be well to consider which series of objectives should be selected for specific work. It has to be remembered that apochromatic objectives are very expensive, and, generally speaking, are beyond the reach of the ordinary amateur, who usually takes up microscopy without special scientific aims, and excepting to a trained critical eye they would not be found to possess the extraordinary merit that is claimed.

The question naturally occurs, Is it worth while to incur the great cost which is involved in the purchase of apochromatic objectives? No decided opinion can be given without a full knowledge of the scope of the work which is to be undertaken, and as from the nature of things it is impossible at its inception to tell the extent to which research may be carried, the difficulty of giving advice is increased. Generally speaking, it may be stated definitely that for the ordinary work of the amateur, the so-called 'students' series' of lenses will be found to give all the pleasure and satisfaction that are to be derived from the examination of Nature's small things, without attempting to obtain apparently impossible results or to detect structure not hitherto discovered. The man who definitely equips himself for original research cannot afford to have less than the very best means which modern optical skill can afford him, and from the point of view of actual supremacy the apochromatics must then be chosen ; but he would be limiting his possibilities in practically no degree whatever by having lenses carefully selected from those previously referred to under the title of semi-apochromatics.

It should be remembered that the reduction of spherical zones enables a higher power of eyepiece to be employed with an objective than would otherwise be possible, and it is due to this quality that the apochromatic objectives have been especially valuable.

A series of compensating eyepieces is specially designed to work with them, having powers varying from 2 to 27 diameters. Supposing, therefore, we were working with a $\frac{1}{4}$-inch apochromat having an initial power of 40 diameters, with a 10-inch tube-length, we could by means of the searcher eyepiece (× 2) obtain a magnification of 80 diameters, and by using intermediate powers of eyepieces up to the × 27, produce any magnification that might be desired from 80 to 1,080 diameters ($\frac{1}{4}$-inch initial power of 40 × 27 eyepiece power = 1,080).

Further, these special eyepieces are all designed to work in the same focal plane at the tube-length for which the eyepieces and objectives are designed, with the result that practically very little refocussing is necessary on the exchange of an eyepiece during an observation.   By this means the magnification with a low-power objective having a long working distance and a fairly high N.A. for its power, as possessed by all of the apochromats of Messrs. Zeiss's manufacture, can be gradually increased, and the advantage gained is one for which many microscopists sighed before the days of apochromats—namely, a wide range of magnifying power and great working distance.   For many classes of work this convenience is very great ; but it must not be forgotten that medium- and low-power eyepieces are desirable for general work, that it is advisable to use such high-power eyepieces for occasional reference or for testing purposes only, and that if high magnification is necessary it is always to be insisted that a suitable high-power objective with a medium- or low-power eyepiece is the only satisfactory means of working.

Some of the best of the achromatic objectives, to which reference has already been made, will stand as high an eyepiece power as the apochromats, but generally they do not advantageously bear anything higher than, say, up to 10 or 12 diameters, and usually not in the perfect manner that the apochromats do. The Huyghenian eyepieces that are used with the achromatics are very rarely designed to work one after the other in the same focal plane, with the result that it is necessary to refocus every time the eyepieces are exchanged, and the higher the power of the eyepiece that is employed the closer will the objective work to the object.

Magnifying power, however, is not the only feature to be considered with regard to an objective ; there must be a power of delineating fine detail. This latter quality is dependent on the numerical aperture of the objective, referred to on p. 62.

IMMERSION OBJECTIVES.—In using these objectives a film of a specified fluid is interposed between the front lens of the objective and the cover-glass of the object under examination, so that continuity is established between them. There are two media that are in regular use, viz., water and cedar-wood oil. Others, including glycerine and mono-bromide of naphthalin, are, however, occasionally employed. It may be taken that when a lens is referred to as a ' homogeneous or immersion ' objective, cedar-wood oil, or a mixture of which that oil is the principal ingredient, known as ' immersion oil,' is the correct medium for using with that objective. The refractive index of cedar-wood oil is about 1·52, and practically the same as crown-glass ; consequently, when it is used for immersion purposes it has the effect of rendering the cover-glass part of the objective.

The question naturally arises, What advantage is gained by the use of an immersion medium ? In reply, it may be briefly stated that the resolving power of the objective, the brilliance of the image, and the working distance, are all increased.

It is a well-known law that rays passing from a rarer to a denser medium are refracted towards the perpendicular, and vice versa. If, therefore, an object be examined with a dry objective, it is obvious that certain rays of light emerging from the denser crown cover-glass into the rarer medium, air, are refracted so far from the perpendicular as to fail to assist in forming the image. By placing a medium between the cover-glass and the objective, these rays are utilized, owing to the influence of the dense medium, oil. The refractive index of air is 1·0, that of water 1·33, while that of cedar-wood oil is 1·52. It will be seen from this that the utility of the oil must be very appreciable, in fact, an oil immersion lens receiving light at 82° and a water immersion lens receiving light at 96° admit the same rays as a dry lens of 180°, and, therefore, divide as many lines to the inch as the maximum number possible with a dry lens. If immersion lenses have greater apertures than the above-named, they will divide finer markings than any dry lens, and they can be theoretically carried to oil and water angles respectively of 180°

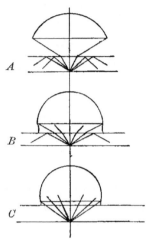

The three diagrams (Fig. 22) will assist in making the matter clearer. They represent the rays of light passing from an object mounted in optical contact with a cover-glass, or in a sufficiently dense medium such as Canada balsam, through the cover-glass.

A shows the front lens of a high-power dry lens over the object. Only the least inclined of the three pairs of rays shown emanating from the object can get out of the cover-glass into the air, the other two being so oblique that they suffer total reflection, and are thus utterly lost to the dry lens.

FIG. 22. — DRY, WATER IMMERSION AND OIL IMMERSION OBJECTIVES, SHOWING THE RAYS RECEIVED BY EACH.

B represents the front of a water immersion objective over the same object. Owing to the refractive index of water being greater than that of air, the first and second pair of rays can now pass through the upper surface of the cover-glass and thus reach the front lens, with a corresponding gain in brightness of the image and resolving power. The most oblique pair of rays is, however, still totally reflected and lost to the water lens.

C shows the front lens of an oil immersion objective. As the refractive index of the immersion oil is the same as that of the cover-glass, all the rays now travel straight on from object to front lens, and none are lost; hence the oil lens gives the brightest image and the highest resolving power.

There is another feature of advantage gained by the use of an oil immersion lens. The refraction caused by the influence of the cover-glass thickness referred to on p. 69 does not take place, on account of the continuity established between the objective and the cover-glass by the immersion oil. There is, therefore, no necessity for such objectives to be provided with a correction collar for variations in thickness of cover-glass; a slight correction of the same kind has, however, sometimes to be made on account of the distance which the object may be beyond the cover-glass when the mounting medium has not the same refractive index as the cover-glass. This can be efficiently effected by either extending or shortening the body-length. Water immersion objectives do not yield so high an aperture as the oil immersions, and as the immersion medium is not of the same density as the cover-glass a correction collar is essential, but there are subjects with which oil could not be suitably used, and in such cases the water immersion lenses have to be chosen.

Mono-bromide of naphthalin is, at present, only used with one form of objective, a $\frac{1}{10}$-inch, by Carl Zeiss, of Jena, having a numerical aperture of 1·63. The refractive index of this medium is 1·657, and special flint cover-glasses of the same density have to be employed with it. This restriction, together with its high price—£40—has prevented its being largely used. Those who have had an opportunity of working with one have spoken in high terms of the beautiful effects it yields.

It must be borne in mind that objectives that are intended to be used immersed are specially corrected for the specific medium to be employed. Ordinary lenses intended for use dry cannot be advantageously worked immersed.

An important point in working with immersion objectives must be mentioned. It has been remarked by workers that the same objective does not always give equally satisfactory effects even with the same object; that definition has gradually improved after a short time in use, and perfectly good objectives have been condemned as poor. The explanation is to be found in the difference or differences of temperature of the object, the immersion oil, and the objective. Occasionally the cover-glass is warm after a specimen has been hastily prepared, and this might produce a very marked effect. The only way, if the parts have not been exposed under the same conditions, is to set up the microscope and let it stand for a short while, so that all may have a mean temperature.

A note may be added on the removal of the immersion medium from the front of the objective. The usual method is to wipe it carefully with a soft handkerchief of silk and cotton mixture or a clean thin chamois leather. This is usually sufficient, but it must be done immediately after use, and with care. Dr. Henri van Heurck always recommended the use of saliva on a piece of old dry linen for the purpose, and stated that, ' thanks to the slight quantity of soda contained in the saliva, the cleaning is perfect and practically instantaneous.'

There are times when, by oversight or force of circumstances, the lens is left uncleaned, and the oil drys on hard and is difficult of removal. The simplest way is, then, to stand the objective on its screwed end, and to put a few drops of the immersion fluid on the dry deposit ; this will, after a short interval, act as a solvent, and the whole will, with a little care and patience, and perhaps a repetition of the process, be easily cleaned off.

APERTURES OF OBJECTIVES—ANGULAR AND NUMERICAL.

For reasons which will be stated hereafter, it will be seen that on the aperture possessed by an objective depends the fineness of detail that it is capable of delineating—that is, the number of lines per inch that it will separate.

ANGULAR APERTURE.—Before the introduction of immersion objectives the ability of an objective to resolve fine structure was known to be dependent on the angle formed by the extreme rays issuing from the object that could be received by the objective. This, which was called the angular aperture of a lens, is, in other words, the angle of the cone which envelops the pencil of light that is received by the objective from a point on the object.

As we have stated in the description of immersion lenses, an oil immersion lens receiving light at 82°, and a water immersion receiving light at 96°, would each divide as many lines to the inch as a dry lens having the limiting aperture of 180° (which in practice can never be quite reached), and as the immersion lenses can theoretically be carried to oil and water angles respectively of 180°, it is obvious that in order to express the efficiency of such objectives a notation must be employed which takes cognizance of the medium which surrounds the front of the objective, and the result it has in the formation of the image. This is achieved by means of the system termed NUMERICAL APERTURE, which was introduced by Professor Abbe. This expresses the efficiency of an objective to allow pencils of light to pass so as to include them in the light forming the image. Numerical aperture is expressed in the formula $n \sin u$. n signifies the index of refraction of the medium by which the objective front is enveloped, and u equals half the angle of aperture. Therefore, by multiplying the sine of the semi-angle of aperture by the refractive index of the medium in which that angle has been measured, the numerical aperture ($n \sin u$) is obtained.

It follows from this that the greatest value which the numerical aperture can have in the case of dry lenses is unity, corresponding to an angular aperture of 180°.

If we are aware of the numerical aperture of an objective we can readily ascertain the number of lines per inch or millimetre which it is capable of dividing, or, in other words, its extreme power of resolution. The formula is—twice the numerical aperture, multiplied by the wave-frequency* of the light used, equals the extreme number of markings per inch or millimetre, according as the calculation may be made, that the lens will resolve.† Conversely, if the extreme limit of resolving power be known, the number of lines per inch or millimetre that it will separate, divided by the wave-frequency of light used, equals twice the numerical aperture.

These calculations are based on the assumption that annular or some other form of oblique illumination is used. With a solid cone of illumination equal to the numerical aperture of the objective no fine detail is visible; it becomes blurred, and, in practice, when using solid cones of illumination, it is usual to

---

* The wave-frequency is the number of waves contained in an inch or millimetre, according to which measure is used.
† The mean wave-length of white light is 0·5269 μ (=48,200 to an inch). Taking the numerical aperture of an objective as 1·0 N.A., and for this purpose doubling it, we find that with the aperture of 1·0 N.A. 96,400 lines (about) per inch can be resolved with white light (48,200 × double the numerical aperture = 2, produces 96,400).

make them fill three-quarters only of the back lens of the objective. This will be found treated on p. 100 in connection with condensers.

Under these conditions the number of lines per inch that will be resolved by the objective will be ascertained by multiplying the wave-frequency of the light used by $\frac{3}{2}$, and then multiplying the product by the numerical aperture.* The foregoing is probably the estimate from practical working, but the theoretical factor is $\frac{7}{4}$.

Measuring Numerical Apertures.

It is always advantageous for the worker to be in a position to measure the N.A. of objectives, and three reliable means of so doing are here given.

The Apertometer.—To enable the numerical apertures of objectives to be taken without a calculation, Professor Abbe devised the apertometer. It consists of an almost semicircular plate of glass, having the diametrical edge ground to an angle of 45°, while the circumference is a polished cylindrical surface. It is shown in Fig. 23.

The centre of the semicircle is marked by a silvered disc, a, having a very small central aperture, and on the upper surface on the periphery it is provided with a scale of divisions, indicating both angular and numerical apertures. The manner of using the apertometer is as follows : The microscope is placed in a vertical position, and the apertometer is laid on the stage, with the diametrical edge towards the limb of the instrument. The objective that it is desired to take the aperture of is then screwed on, and the objective focussed on the plain centre of the silvered disc. It is well now to fix the apertometer to the stage, either by springs or an elastic band, to prevent its moving. The two pointers, b, are then set on the edge of the circle to read zero. The draw-tube and eyepiece with which the silvered disc has been set are removed, and at the lower end of the draw-tube a special objective of low power, that is supplied with the apparatus, is screwed into the universal thread, which should be there fitted in all microscopes of high class.

The cylindrical edge of the apertometer is then illuminated in front and at the sides by means of bull's-eye condensers and

* E. M. Nelson, *Journal of the Royal Microscopical Society*, 1893, p. 15

lamps, or, if daylight is available, it will be easier to get uniform brilliance all over the field by placing the microscope on a table in front of a window, and using bull's-eye condensers to increase the light.

The draw-tube carrying the special objective and the eyepiece is then replaced in the body of the microscope, and the image of the pointers *b* in Fig. 23 is brought sharply into focus in the centre of the field by slowly extending the draw-tube, being reflected by means of the prismatic diametrical edge of the glass plate. While looking through the microscope these pointers are then each moved separately, in opposite directions, round the

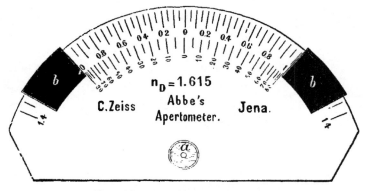

FIG. 23.—ABBE'S APERTOMETER.

outer edge of the apertometer, until they are set exactly on the extreme margins of the back lens. The reading is then taken by the divisions on the face of the apertometer against which the pointers have now arrived. In the case of an oil or water immersion lens, the medium must, of course, be placed in front of the object-glass during the examination.

Caution is necessary in applying this instrument to some of the modern objectives with unusually large lenses, and also in applying it to condensers. The reason for this is that the auxiliary objective supplied with the apertometer, which is to be screwed to the end of the draw-tube in order to produce a microscope for clearly observing the aperture of the objective under test, is of too small a clear diameter to receive all the light which objectives and condensers of the kind mentioned are capable of passing; the result being that the N.A. obtained under such conditions is too small.   It is necessary in such cases to do away with the auxiliary objective altogether, and to observe the back of the objective under test by looking directly at it down the microscope-tube without the intervention of any lenses whatever.   With objectives of such a large diameter as those to which this warning applies there is no difficulty whatever in obtaining an accurate setting by naked-eye observations.

CHESHIRE'S APERTOMETER.—This is quite a simple device, and enables a sufficiently exact estimate of the numerical aperture to

FIG. 24.—CHESHIRE'S APERTOMETER, BY R. AND J. BECK.

be obtained for most purposes.   Familiarity with it renders it more reliable than would be expected at first sight.

The instrument as made by R. and J. Beck consists of a circular disc of glass with a mark on the upper surface to which the object-glass being tested is focussed.   The lower surface is ruled with a series of concentric rings, each of which is placed so as to correspond to 0·1 N.A.

The method of use is to place the apertometer on the stage of the microscope, focus the cross-lines ruled on the upper surface, then remove the eyepiece and count the number of rings which show through the back lens of the object-glass.   For high powers, when the lens is small, and consequently the rings are difficult to count, a special eyepiece which focusses to the back focal plane of the object-glass is supplied, which is inserted in place of the usual eyepiece.

MR. CONRADY'S METHOD FOR DRY OBJECTIVES.—In the case of dry objectives the numerical aperture may be determined without expensive apparatus of any kind, and with equal accuracy to that obtainable with the Zeiss apertometer, by any method suitable for determining the angle of the objective.

Perhaps the easiest way of doing so is the following:

On a dark background, such as a table, place two white cards with their inner edges parallel to each other and a suitable distance apart.   If a dry objective is held at a sufficient distance above the table and directed towards a point midway between the two cards, an image of the latter will be seen at the back of the objective, and by approaching the objective to the table these images will recede from one another, until finally they can be got to disappear under the margin of the back lens.

It is obvious that when this disappearance takes place the inner edges of the two cards are lying in the direction of the most oblique rays which can enter the objective. In other words, they form, with the focal point of the objective, its angle of aperture.

In order to get numerical results, the two cards must be placed at a measured distance apart, and the objective made to slide up and down along the edge of a divided scale, such as an ordinary foot rule.

When the point of disappearance of the images of the inner edges of the cards has been reached, the distance from the front of the objective to the table is read off on the scale. This is further diminished by the working distance of the objective, which must be separately measured, and the numerical aperture is then obtained as follows :

1. If a table of trigonometrical functions is available, divide the reduced distance from objective to table by half the distance between the inner edges of the two white cards. The quotient is the co-tangent of half the angle of aperture ; find its value in the table, and take from the table the sine of the same angle. This is the desired numerical aperture.

2. If no trigonometrical table is available, the numerical aperture is found as follows :

Square the reduced distance from objective to table, also half the distance between the two cards. Add the two squares together and extract the square root of the result. Then the numerical aperture is found as the quotient of half the distance between the cardboards, and the value of the above square root.

In order to make the calculation involved as simple as possible, it is manifestly an advantage to make one of the lengths entering into the calculation unity, which is easily done by placing the two cards a distance of two units apart.

For objectives of low numerical aperture 2 inches will be found a suitable distance ; for those of higher numerical aperture a distance of 2 decimetres. This last measurement is suggested so that the experiment may be in its simplest form.

NUMERICAL APERTURE AND POWER.—As we have before remarked

magnifying power is not the only quality necessary for the observation of minute structure. The power to delineate fine detail is still more dependent on the numerical aperture of the objective. It has been explained by Dr. Dallinger in ' Carpenter on the Microscope,' and will be evident from a consideration of the preceding remarks concerning numerical aperture, that two objectives—one of much greater magnifying power than the other, but both having only the same numerical aperture—will only divide the same amount of detail, the higher power exhibiting it on a larger scale. That is, supposing with a $\frac{1}{4}$-inch objective of 0·90 N.A. certain structure were presented, and then a $\frac{1}{8}$-inch objective with just double the magnification, but with the same N.A., were afterwards used, there would be no further power of resolution in the $\frac{1}{8}$ than in the $\frac{1}{4}$. It might be possible to make an objective of very low power of sufficiently high aperture to divide very minute details, but this would be useless unless the objective would bear a sufficiently deep eyepiece to enable the human eye to see it. It therefore becomes necessary that a ratio of aperture to power should be established. Mr. Nelson has suggested that a standard, termed the ' optical index' (O.I.), should be adopted for this purpose, to indicate the numerical aperture that should be given to an objective, if it be intended that the eye should see in the image as fine detail as it could divide in a real object of the same size. It is ascertained by multiplying the numerical aperture of an objective by 1,000, and dividing by the initial magnifying power of the objective.* If a microscope is required to show all that keen eyes are able to appreciate, then 0·26 N.A. must be given to it for every 100 diameters of magnification. If we limit the power of the eyepiece of such a microscope to 10, then the objective must have 0·26 N.A. for each 10 diameters of *initial* magnifying power. The optical index, therefore, of an objective which, with an eyepiece magnifying 10 diameters, will yield all that it is possible for a normal eye to appreciate, will be 26·0. In practice it is found possible to employ eyepieces giving higher magnifications than those mentioned in Mr. Nelson's rule, and these would, of course, enlarge the image.

* E. M. Nelson, *Journal of the Royal Microscopical Society*, February, 1893.

Although large apertures are the pride of those whose ultimate ambition in matters microscopical seems to be bounded by the endeavour to resolve the markings upon diatomaceous frustules, t is doubtful whether for the ordinary amateur there is a real necessity for the extremely large apertures. Lenses having such, require great skill and care in manufacturing and adjusting, and are consequently expensive; and if the ordinary work of an amateur is proposed to be conducted, and not original scientific research, objectives of medium aperture will usually be found to meet his requirements thoroughly.

THE INFLUENCE OF THE COVER-GLASS.

As a rule, objectives are corrected for a specified thickness of cover-glass, which is placed over the object to protect it. These cover-glasses, however, vary considerably in thickness, and consequently by refraction disturb the corrections of the objectives. An objective which gives crisp definition when an object that has no cover-glass to it is being viewed, will not define so clearly if a thin one be applied, and the greater the thickness of the cover-glass the more will the image be deteriorated.

In other words, spherical aberration in the sense of under-correction is introduced when the cover-glass is thinner, and in the sense of over-correction when the cover-glass is thicker, than that for which the lens was adjusted. This spherical aberration arises from the refraction of the rays in the plane surfaces of the cover-glass and objective front respectively; it is negative for the cover-glass surface, and positive in the front lens plano, the latter preponderating owing to the greater diameter of the cone of rays from the object where it enters the front lens. With the correct thickness of cover-glass, the remaining under-correction is exactly balanced by an equal over-correction in the lenses of the objective, but a thin cover-glass produces insufficient over-correction, the diameter of the cone of rays being too small when it meets the surface of the cover-glass. A thick cover-glass produces the opposite effect—that is, the cone of rays is too large.

Low powers are not so sensitive to this influence as high ones. There are two means of correcting this. Dry objectives having a large numerical aperture are often provided with what is termed 'a correction collar,' whereby the two back combinations of the objective are removed farther from, or brought closer to, the front lens or lenses. Fig. 25 shows the manner in which this is effected in one of Zeiss's lenses : bb is the correction ring, by turning which the distance between the upper lenses and the two lower lenses is varied. With such a correction collar a worker is undisturbed by thickness of cover-glass, because he has within certain limits the means at his disposal in the objective itself for correcting same. The use of this correction collar almost requires a personal demonstration, and to set it at the exact point that yields the best result is a matter of extreme delicacy, which can only be accurately done as the result of experience and with the aid of a critical eye. It is hardly to be recommended to students, because they will not usually afford the time and trouble necessary to get such perfect results; consequently there is a growing tendency, except in the apochromatic objectives, to have the lenses mounted in a rigid setting, corrected for a specific tube-length and thickness of cover-glass. With the fixed

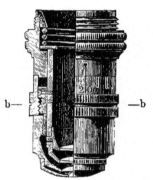

FIG. 25.—CORRECTION COLLAR (ZEISS).

setting, if a different thickness of cover-glass be used than that for which the objective was designed, correction can be made by altering the tube-length of the microscope. This has the same effect as altering the distance between the lenses. Supposing we had an objective adjusted for a 6-inch tube, with a 'B' eyepiece, on a cover-glass 0·008 inch thick (this is about the average thickness adopted by opticians), and we wished to examine an object having another thickness of cover—say 0·005—we should at once notice that the performance was not so good, and in order to improve it we should have to make the body longer. This difference of cover-glass thickness, with a good ⅙-inch objective, would necessitate a 10-inch body. On the other hand, if the cover were 0·01 inch thick, the length of the body would have to be diminished below that for which the object had been designed, to obtain the best results—that is, for a thin cover the body-tube would have to be lengthened, while for a thick one it would have to be shortened, and the finer the quality of the objective the more sensitive would it be to cover-glass thickness. This system of correcting by draw-tube, however, has one drawback, and that is, that the power is varied in correcting, and, of course, the focus is altered. From the considerations here named, it will be found advantageous if the microscope be provided with a means of lengthening the body by draw-tubes to 12 inches, and on the other hand, when the draw-tubes are closed, of having the body shorter than the Continental length (6 inches). In order that the best adjustment may be made, it is essential that one of the draw-tubes be actuated by rack and pinion, and the convenience of this arrangement cannot be too strongly urged upon microscopists. Messrs. Baker, Beck, and Watson and Sons have adopted it in their large models, and it has evidently met with considerable appreciation.

DIRECTIONS FOR USING A CORRECTION COLLAR AND CORRECTING BY TUBE-LENGTH.

This may be accomplished in a systematic manner if it be borne in mind that the aim is to eliminate spherical aberration, which defect may be defined as a difference of focus between the central and marginal zones of an objective. Hence the correct tube-length or the best position of the correction collar has been found when some strongly marked detail or outline of the object remains in exact focus under *any* change of illumination, say from a small to a large diaphragm opening beneath the condenser, or, better still, by changing the illumination from central to very oblique, these changes being made with great care, so as not to disturb the other adjustments.

The following process will be the safest and quickest: Start with the shortest tube-length, or when there is a correction collar, with the position corresponding to the thickest cover-glass; carefully focus some sharp outline with, say, a $\frac{1}{4}$ central cone, then change to a $\frac{3}{4}$ cone, or, better still, to very oblique light. Unless the object—owing to an exceptionally thick cover-glass, or a very badly adjusted lens—is beyond the range of your adjustments, you will find evidence of under-correction—that is, the lens will have to be brought closer to the object with the wide cone, or oblique light, than with central light.

Gradually lengthen the tube, or turn the collar, repeating the above observation after each change, until all evidence of spherical aberration has disappeared; the instrument is then in correct adjustment within your own limits of vision.

It is advisable to start with the adjustment corresponding to the thickest cover, for the simple reason that this lessens the danger of running through the cover-glass and destroying the object, and possibly the front lens of the objective, when dealing with a lens of a short working distance.

The difference between an objective adapted to a 6-inch and that for a 10-inch tube is, that in the latter case the back combinations of the objective are brought closer to the front lenses. This gives a slightly increased aperture. The majority of cover-glasses that are purchased and a large number of those used over commercial objects are more than 0·007 inch thick; 0·007 inch is a medium thickness of cover-glass, but the tendency is to use thicker ones. It will be found a great advantage to buy only such objectives as are corrected for a medium tube-length,* and having the rackwork before referred to fitted to the microscope-tube, sufficient latitude would still be allowed if a thinner cover-glass were met with; but it would often be found necessary to close the draw-tubes down to 6 or 7 inches, in order to get the best correction for the thick cover-glasses that are commonly used.

We may here clear up another question that occasionally arises. If a $\frac{1}{6}$-inch objective is corrected for a 6-inch tube-length, it does not give a magnification of 60 diameters at 6 inches. The powers of all objectives are calculated for a 10-inch tube-length, therefore the full total benefit is not obtained from an objective when used at 6 inches, but only six-tenths of it. Of course, with the lessened magnification at 6 inches a brighter field is produced, and a deeper power of eyepiece is found permissible. This is rather an important item in testing an objective, because an objective at 10 inches would be yielding about two-thirds more magnification than at 6 inches, and its

---

* Mr. Conrady has advocated that all objectives should be corrected for a tube-length of 8 inches, and with excellent reason, for such an arrangement would be a practical step in the solution of a difficult problem.

powers would be much more severely tested than if employed at 6 inches.

It would be a great advantage to the microscopists if opticians would mark exactly the focal power and precise numerical aperture of their objectives upon them. In order that objectives may appear to have a large ratio of aperture to power, they are often put forward as possessing a considerably lower power than they actually have. For instance, a so-called 1-inch often turns out to be nearer $\frac{2}{3}$-inch, $\frac{1}{2}$-inch about $\frac{4}{10}$-inch, $\frac{1}{5}$-inch to have the power of $\frac{1}{6}$-inch, and $\frac{1}{12}$-inch in some instances to be $\frac{1}{14}$-inch. It has become such an acknowledged fact that the act of misrepresentation involved seems to be condoned. This is a state of things which should not be. Opticians must be aware of the misdescription and the immensity of trouble that is caused by it. We must, therefore, advise microscopists not to rely on the powers marked on their objectives, but to ascertain them for themselves, and the best way to do it is to project the image of a micrometer, without any eyepiece in the body-tube, on a screen 20 inches distant from the back lens of the objective. Measure with a foot rule the distance apart of the lines so projected, and supposing that each hundredth of an inch measured on the screen 1 inch, that would represent a magnification of 100 diameters; divide the distance used (20 inches) by the magnification found (100 diameters), and the result ($\frac{20}{100}$ or $\frac{1}{5}$-inch) is the equivalent focus or 'power' of the objective.

TESTING OBJECTIVES.

It is a somewhat difficult matter for the novice to decide for himself as to the quality of object-glasses. Such work needs experience, judgment, and a trained eye. The writer has met with people who have not been able to distinguish the difference in performance between an uncorrected single French lens and a first-class achromatic. This, of course, was due entirely to a lack of that perception of microscopical detail which can only be acquired by intimacy with objectives and their qualities. Especially is this true in lenses of the highest grade. We propose, therefore, to give a few hints which, if not of so much use in the initial stage, may be of aid at a later period.

FLATNESS OF FIELD.—This feature, however much it may be appreciated, and greatly as it is to be desired, is, unfortunately, impossible of association with objectives of fine quality. With low powers up to ½ inch it is generally obtainable for a considerable portion of the field, especially with objectives of small numerical aperture. The compromise which is to be made to secure it is not such in the low powers as to materially affect the general performance. It cannot, however, be given in objectives of medium and high power. A well-corrected objective inevitably has a curved field, and the more perfectly it is corrected, the more apparent does it become. This has become increasingly recognized, and it is now conceded that it is better to get the utmost perfection of definition in the central zone rather than that sharpness should be sacrificed to flatness of field. Flatness of field in any other than low-power objectives cannot therefore be expected except at the expense of inferior definition, and this to the critical worker would be intolerable.

It must be noted, however, that every part of the field can be separately brought into view with all well-corrected lenses by slightly altering the focus with the fine adjustment.

COLOUR.—Dr. Carpenter's old test for achromatism—the examination of the cells in a thin section of deal—will give a very good idea of the colour corrections of objectives. For high powers, the markings on a frustule of the diatom *Pleurosigma formosum* are an excellent test. With the apochromatic objectives these come out quite black and white, while with those of the achromatic series any outstanding colour is at once revealed. Another method is the mercury test adopted by opticians. A small globule of mercury is placed on a slip of ebonite, and a piece of whalebone or watch-spring is made to snap on it, causing the globule to split up into numerous particles of exceedingly minute size. These globules are then examined with the objective, and can be illuminated by means of a bare gas-jet, lamp, or daylight. Outstanding colour will be revealed by the globules. A cover-glass of proper thickness must be interposed when submitting lenses of considerable numerical aperture to these tests.

The Abbe Test-Plate.

The most satisfactory way of testing an objective that is at the disposal of him who would learn the whole inwardness of his lens is the Abbe test-plate. A considerable amount of experience will be required to use it advantageously, but it discloses at once, when employed by one who has learnt to appreciate its significance, any mechanical inaccuracies that may exist. Directions for using accompany each plate, but the worker will quickly mark out for himself a line which experience will show him is the best for ascertaining whether the lens is accurately centred and the state of the corrections for spherical and chromatic aberrations. The test-plate itself consists of six discs of cover-

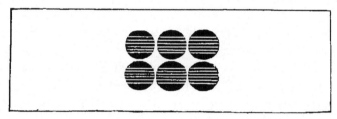

FIG. 26.—ABBE TEST-PLATE.

glass, all of different specified thicknesses, and embracing such a range as objectives are likely to be corrected for.

On the under surface, lines are ruled in a deposit of silver, and the covers are mounted on an ordinary 3 × 1 slip. The ruled lines are coarse, and can be separated with a low-power objective. The procedure adopted by the writer is as follows:

The tube-length should be that for which the objective is ostensibly corrected. An eyepiece of high power and a sub-stage condenser, giving a solid cone equal to at least two-thirds the total aperture of the objective, are used.

In this connection it may be mentioned that it will be found advantageous to have an eyepiece which permits of some degree of over- or under-correction being obtained—such, for instance, as will be afforded by the Holoscopic eyepiece described on p. 85. Those who have not such a convenience may unscrew the eye lens of the Huyghenian eyepiece and so secure some slight modification of correction. The object of this is to ascertain whether the lines may be rendered free from coloured edges or with the same colour on both edges of all the lines in the field. If in this preliminary step it be found that the definition is unsatisfactory, thicker and thinner cover-glasses should be tried ; and in the event of failure to secure good definition in this way, and no reasonable alteration of tube-length will produce the desired effect, the objective may be safely rejected as bad.

It probably will be found that under one of the cover-glasses the lines will appear satisfactorily defined, in which case the centring may be examined.

DEFECTIVE CENTRING shows itself (a) by the impossibility of removing the coloured edges of the lines all over the field even when the eyepiece is adjusted as described above, the edge colouring being more apparent on one side of the field than the other ; (b) by unequal definition of the two edges of the central lines, one edge appearing sharp or nearly so, while the other edge is seen double or foggy.

SPHERICAL ABERRATION.—The fact that the objective will bear high-power eyepieces on the test-plate in a satisfactory manner is in itself proof of good correction in this respect, but the following is a further excellent and convincing test :

Place a diaphragm beneath the condenser having an aperture that will cause the condenser to yield a cone of illumination equal to one-fourth the N.A. of the objective under examination, and while observing the lines change the position of this diaphragm from central to extremely oblique, the obliquity being in a plane at right angles to the direction of the lines. This is best performed by means of one of the mechanical condenser carriers, such as are provided in the Continental microscopes. If the lines remain sharp throughout, the corrections for spherical aberration are eminently satisfactory ; but should a difference of focus occur, to avoid all chances of erroneous deduction the other discs should be examined to insure that the proper thickness of cover is being used ; also the tube-length might be varied, and if after these precautions it is still found that there is a difference of focus over the intermediate position of the diaphragm, the existence of a spherical zone is at once demonstrated. This process enables the best tube-length and thickness of cover for the objective to be discovered with accuracy.

CHROMATIC CORRECTION.—Tests for this should be made in the same manner as for the spherical correction. Under the same conditions an apochromatic should show practically no colour, or, at the most, barely distinguishable traces of tertiary tints. Semi-apochromatics, or lenses of fine correction, will show narrow bands of pale green (apple-green) on one side, and faint purple (or claret) on the other side, of each line, and the same colours or tints should appear whether the diaphragm be used centrally or obliquely, the width of the colour bands only changing; further, good definition should be yielded under all circumstances. Ordinary lenses will generally show the best colour correction for some intermediate zone. If they exhibit broad bands of primary colour—yellow or blue—with very oblique light, the definition will be found to be bad.

CURVATURE OF FIELD.—There is one point concerning which some slight difficulty may arise in connection with the curvature of the field. Absence of flatness of field, which is inherent in the construction of all latter-day objectives, and particularly so in those of high power and large aperture, is not regarded as a fault, but coma in an objective at first sight gives the same appearance; the difference, however, is this: when the latter fault exists only the central line or those nearest to it can be focussed sharply, those towards the margin remaining indistinct or ill-defined when refocussing is attempted; in a well-corrected lens all of the lines will become sharp and distinct when the particular zone is adjusted for.

It will be found advantageous to confirm the observations on the test-plate by examinations of known test objects, and with practice the two together will soon enable reliable estimates to be formed of the quality of objectives.

TESTS FOR DEFINITION.—Use an eyepiece with the objective under examination that will give a total magnification in diameters equal to one thousand times the numerical aperture of the objective—that is, if a $\frac{1}{2}$-inch objective, having a magnifying power of 20 diameters on the 10-inch tube, had a numerical aperture of 0·45, an eyepiece, the magnification of which was $22\frac{1}{2}$ diameters, would be necessary to give the required 450 diameters. If an objective bears this without serious breaking down its definition may be considered to be good. This test has the advantage of being based on a rational foundation, the ratio being the same as an eyepiece power of 50 to each inch of aperture in an ordinary telescope. This, again, is equal to what would be seen of an object if looked at through a pin-hole $\frac{1}{50}$ inch in diameter, beyond which the outlines of objects fail in clearness.

For objectives varying in power from 2 inches to $\frac{1}{2}$ inch, nothing is better as a test than the proboscis of a blowfly. The spines in the central portion of the tongue should each show a well-defined point. For high-power objectives the internal markings of Triceratum and *Pleurosigma angulatum*, also the markings on the scales of Podura (*Lepidocyrtus curvicollis*), are the most suitable.

CHOICE OF OBJECTIVES.

The best objectives for a novice at starting would be 2-inch, 1-inch, and $\frac{1}{6}$-inch. The 2-inch will be found extremely useful for large specimens, while the 1-inch, which is considered the working-glass of the average microscopist, will with a higher power—namely, the $\frac{1}{6}$-inch—show him some of the minuter detail which sooner or later he will wish to make himself acquainted with. If more object-glasses than these be required, we should recommend the $\frac{1}{2}$-inch as an intermediate between the 1-inch and the $\frac{1}{6}$-inch, and for a higher power a $\frac{1}{12}$-inch oil immersion objective should be added. For medical work the $\frac{2}{3}$-inch is invariably chosen with the $\frac{1}{6}$-inch. It is well to buy only such low-power objectives as have double combinations. Some of the cheaper ones consist of two or three lenses balsamed together in one combination only; with these there is an insufficiency of aperture, and good definition and flatness of field cannot be obtained. All the best low-power lenses are constructed with two pairs or more of lenses set a little distance apart, and can be readily recognized. Of the apochromatic series of Zeiss our choice would be the 24, 12, 6, and 2 millimetre objectives if for the English tube-length, or the 16, 4, and 2 if for the Continental tube-length. Whether the 2-millimetre objective should be that with the maximum numerical aperture of 1·40 or that having 1·30 depends on the work to be done, and the worker. The lens of lower aperture is less liable to injury than the other, and is consequently much more largely used. Generally speaking, 1·30 will be found sufficient, but that of 1·40 will be the lens for the worker who requires the utmost capacity in his lens, and will give it due care. A ⅛-inch with the remarkable working distance of 1 millimetre has been introduced by Watson and Sons, and is excellent in performance, being of the semi-apochromatic variety. Swift and Sons' pan-aplanatic lenses are deservedly popular, while the objectives of Leitz and Zeiss are uniformly good. As previously remarked, competition has caused all the makers to bring their objectives to a high level of perfection, and the novice will be quite safe in equipping himself with those of any of the leading English opticians. The productions of American opticians are but little known in England, yet they are said in many instances by competent judges to be of exceptional quality, Messrs. Bausch and Lomb and the Spencer Lens Company having high reputation for their lenses.

Of course, different requirements would necessitate the selection of special objectives, but a practical microscopist or any microscope manufacturer would be able to advise on the matter.

HOW A MICROSCOPE OBJECTIVE IS MADE.

The knowledge of the processes involved in the manufacture of micro-objectives is so vague that a brief description will be of interest to most readers.*

The mathematician so computes an objective that, when it is constructed from the optical glasses which he has chosen as suitable for their refractive and dispersive qualities, with the component lenses shaped to the thickness and curvature he has prescribed, and finally mounted and accurately centred at the computed distances, it will exactly realize his intentions.

In optical glass he has a wide choice, but when a melting is exhausted it rarely happens that an exact duplicate can be obtained from the glass manufacturers, two meltings seldom agreeing with sufficient precision in their refractive and dispersive

* This description was given at a demonstration given at the Quekett Club by Mr. F. W. Watson Baker, F.R.M.S.

powers. It therefore becomes necessary to re-compute the lens for each change of its constituents.

Incidentally, there is here an explanation of the reason why the replacing of any portion of an objective system years after its manufacture rarely gives the same results as the original.

Convex and concave templets or gauges are then made of the radii of the various constituent parts; and pairs of tools are next turned to fit the curvature of the gauges. These tools are divided into three classes—

Roughers, True Tools, and Polishers.

They are similar to A and C in diagram below, and are fixed to a rotating spindle similar to that of a lathe, which, however,

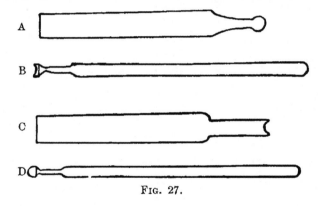

Fig. 27.

can be either vertical or horizontal. These tools must of necessity be accurately curved to the spherical shape that the glass is to assume in lens form.

THE GLASS.—The glass is received in thick slabs or plates, and by means of a slitting machine, consisting of a rapidly rotating iron plate charged with diamond dust and oil, thin plates are cut off to approximately the thickness of the lens.

PREPARATION AND PROCESSES.—These thin plates are then cut into small pieces by means of a diamond.

Hand-shanks are next used to take off the square edges and to render the small piece of glass nearly round.

It is then cemented to a holder which rotates in the lathe, and by means of a sharp tool of steel and water is edged to within a fraction of its ultimate diameter.

Then, with the same tool or sometimes with a diamond, the face is shaped spherically.

The lens in its rough state is now removed in its holder from the lathe, and the roughing tool takes its place. The lens fixed to its holder as in B and D is now held in the hand ; the lathe is rotated, and by means of emery, moistened with water, the lens is ground against its corresponding tool A or C with a peculiar rotary movement of the wrist, and as the shape of the lens becomes more true, a finer and yet finer emery is used until the 'figure'—that is, the ultimate curve—is put upon it ready for polishing.

FOR POLISHING, a tool lined with a composition consisting largely of pitch, and moulded true to curve, is fixed and rotated in the lathe, and the glass is continuously worked with it, special polishing materials being used, such as rouge or putty-powder ; and from time to time, at the figuring stage and during the polishing, the curve is tested by means of a 'proof plate.'

THE PROOF PLATE.—This proof plate consists of a plate of glass which has been worked so as to precisely fit the exact curve of the lens which it is intended to test by its means. The accuracy of the proof plates is ascertained by micrometer gauge and spherometer.

If a carefully cleaned lens be brought into contact with the curve in the proof plate and it is nearly correct, then the phenomena known as 'Newton's rings' will appear. These coloured rings are produced whenever two reflecting surfaces are brought very close together, for they are due to the interference of the light reflected from one of the surfaces with that from the other, and the beautiful colours seen in soap-bubbles are a well-known example of them. These rings form an extremely delicate test for the truth of lens surfaces ; for, roughly speaking, the colours run through the complete range of the spectrum for every increase of the space between the adjoining surfaces of the lens and proof plate by one fifty-thousandth part of an inch ; and, as even moderate changes of tint in any one colour are easily perceptible, it is easy to see that by this test irregularities in the surfaces of so little as one-millionth part of an inch can be detected. When a lens is absolutely correct to the proof plate, the appearance is either that of one uniform tint of colour over the entire surface, or, if the contact is a little closer on one edge than on the other, then straight bands of colour will appear. Any difference of curvature between the lens and proof plate betrays itself by the appearance of rings, and if the lens surface is not truly spherical the rings become deformed from the true circular shape which they should show, being elliptical or sometimes even triangular in form.

FINAL STEPS.— When the surfaces of the lenses have thus been figured true to the proof plates, the lens is once more

FIG. 28.—PROOF PLATE AND NEWTON'S RINGS.

mounted on the lathe-spindle to be centred—that is, to have a smooth edge turned and ground, perfectly true with the optical axis of the lens.

The various constituents are then cemented together and baked for several hours, and are subsequently mounted in their brass fittings, and the adjustment for axial truth and distance completes the process.

EYEPIECES.

The eyepiece commonly used with the microscope is what is termed the Huyghenian form, which generally consists of two plano-convex lenses placed at a distance apart about equal to half the sum of their foci, with a stop in the principal focus of the eye-lens. This will be found to meet all ordinary requirements of microscopical work with achromatic objectives. Eyepieces vary in power, and these powers are usually designated by the letters A, B, C, D, etc., A being the weakest power. On the Continent they are generally designated 1, 2, 3, 4, etc., while some firms express their power in focal units; for instance, an eyepiece having a power of 10 would be 1 inch. This last method, or that adopted by Zeiss for the compensating eye-pieces, and several progressive English houses for their ordinary eyepieces, where the actual magnifying power is engraved on the cap of the eyepiece, is the only rational one. The letters A, B, C, etc., or Nos. 1, 2, 3, etc., convey no real idea of the magnifying powers of the eyepieces, because each maker has his own formula for each eyepiece, and there is no correspondence in the powers of one eyepiece marked ' D ' by one maker and that supplied by another. It is often remarked that the Continental objectives stand a stronger power of eyepiece than the English, and on this account a superiority has been claimed for them ; but it should be borne in mind that English manufacturers give in many instances as deep a power of eyepiece as 20 or 25, whereas Continental manufacturers rarely supply them of greater magnifying power than 10 or 12 diameters. In the English series the variation in power between two consecutive eyepieces is generally greater than in the Continental series. A comparison, therefore, between the merits of an English object-glass tested with, say, a ' D ' or No. 4 English eyepiece and a Continental object-glass of the same power tested with a No. 4 eyepiece

FIG. 29.—HUYGHENIAN EYEPIECE.

of Continental make would not be fair, as the former, having a deeper power eyepiece on it, would be liable not to give such perfect results as the latter. There is no reason why a standard series of eyepieces should not be established, to which all makers could conform ; it would greatly add to convenience in working. Although people very often buy deep-power eyepieces, it is advisable, with ordinary achromatic lenses, that no stronger power should be used than an eyepiece giving an initial power of 10 or 12 diameters. The best eyepiece for general purposes is the ' B.' This gives a convenient size of field, and is by far the most comfortable to work with of the whole series. Next, and in addition to this, we should recommend either the ' C ' or the ' D.'

Microscopists having abnormal vision, and preferring to work without spectacles, should have an auxiliary cap made to fit over their eyepieces, carrying a lens of the power that corrects the error of vision. This is especially necessary where measuring has to be done, or where the microscope is arranged for a second person's inspection.

At times it is desired to know whether an eyepiece can have its diaphragm enlarged so as to give a larger field. An easy method of ascertaining how much of the field lens is employed is to make a spot with ink near the margin on the convex side of the field lens, and on placing the eyepiece in the microscope, if the diaphragm has a sufficiently large aperture, the ink will be visible ; if not, it may be enlarged until it appears. The diaphragm should not be so large as to admit of more than the edge of the field lens being visible.

Compensating Eyepieces.

Under the description of 'Apochromatic Objectives' on p. 55, reference is made to compensating eyepieces. These are specially designed to correct an outstanding colour defect (of the nature of under-correction) which is inherent in all high-power objectives, whether they be apochromatic or achromatic, on account of the peculiar construction of the front lens. For the sake of uniformity of eyepiece, Zeiss imparts the same colour effect to the lower-power lenses of the apochromatic series. The eyepieces, then, have an equal error of the opposite kind (over-correction), and when the objective and eyepiece are combined, a perfect correction is obtained.

The apochromatic objectives are 'under-corrected,' while the achromatic objectives of low power are 'over-corrected.' The compensating eyepieces for the former are over-corrected, and the Huyghenian eyepieces for the latter under-corrected. With low powers of the achromatic type, the compensating eyepieces are disadvantageous ; but with high powers, where the defects caused by the hemispherical front lens give rise to error identical with that in the apochromatic objectives, and which the compensating eyepieces are designed to overcome, these special eyepieces can be employed, but the result is not sufficiently beneficial to justify the purchase of them for use with high-power achromatic objectives only.

A most advantageous feature is imparted to the Zeiss compensating eyepieces. They are all designed to work in the same focal plane, so that when two eyepieces of this series of different powers are interchanged in the body of the microscope no alteration in the focussing is necessary.

A number is engraved on each eyepiece, which, multiplied by the initial magnifying power of the objective, will, when used at the tube-length for which the eyepiece is designed, indicate the magnifying power that is being employed.

The compensating eyepieces designed for the 6-inch tube-length can be used on the 10-inch tube, and those for the 10-inch at the 6-inch tube-length, without detrimental effect ; but in the former case about half must be added to the product of the multiplication of the power of the objective and eyepiece, and in the latter case about one-third must be deducted, in order to arrive at the magnifying power (see *ante*, p. 55).

Holoscopic and Universal Eyepieces.

These eyepieces are made by J. Swift and Son and Watson and Sons, and are intended to be used with objectives of both the apochromatic and achromatic types. The lenses used are made of a selected optical glass, which produces a degree of over-correction similar to that associated with compensating eyepieces when the separation between the eye and field lenses is increased. In order that this may be conveniently effected the eye-lens is attached to an inner, or draw-tube, sliding inside the outer tube which fits the microscope body. When the eyepiece is closed together it becomes of the Huyghenian type ; when the eye-lens tube is pulled out, it gives the effect of a compensating eyepiece. The amount of over-correction can be exactly

FIG. 30.—HOLOSCOPIC EYEPIECE.

obtained by the greater or lesser extension of its draw-tube, a scale being provided to record results. This eyepiece is, therefore, applicable to all classes of objectives, and, being made in a useful range of magnifications, will be found a desirable pattern to start with, and acquaintance with its working will lead to greater appreciation of it. It yields excellent effects photographically. Even if it be intended to limit the equipment to achromatic objectives, this type of eyepiece will generally be found to present points of superiority over the ordinary Huyghenian pattern.

Kellner Eyepieces.

This is an achromatic form of eyepiece, giving an exceedingly large field, which is considerably used for the examination of animalculæ, pond life, etc. A certain amount of definition is, however, sacrificed in working with it; and although occasionally of use, we should recommend the microscopist, before purchasing, to judge for himself as to the desirability or otherwise of his having them. They are not by any means necessary adjuncts.

Projection Eyepieces.

These were designed specially for projecting objects on a screen and for photographic purposes. They give an exceedingly small field, but an exquisitely sharp one. In order to obtain good results with them, it is necessary to alter the position of the eye-lens until the image of the diaphragm appears sharply projected upon the screen. For this purpose the eye-lens is mounted in a tube having in it a spiral slot, permitting of the eye-lens being moved to and fro with great precision. They are usually made in four powers, magnifying 2 and 4 diameters respectively for the short tube-length, and 3 and 6 diameters for the English tube-length. The most serviceable are the 4 for the short tube and the 3 for the long tube. All photo-micrographers of note use these eyepieces, and they can usually be employed to advantage, even with low-power achromatic objectives of good quality. For photographing with ordinary objectives of low power, the ' A ' eyepiece gives good results.

FIG. 31.—PROJECTION EYEPIECE.

Binocular Eyepieces.

As mentioned under 'Binocular Microscopes,' neither of the leading Continental firms make the Wenham binocular microscope. For those, therefore, who desire to be able to employ both eyes, they make a binocular eyepiece, suitable for both high and low powers, but only for the Continental length of tube, and this should be particularly understood. The one with which we are acquainted is that designed by Abbe and manufactured by Zeiss, as shown in Fig. 32. In some hands it gives very beautiful results, while other workers have failed to derive advantage from it. It is designed to give stereoscopic effects and to work with both high and low powers. If it were mounted in some lighter manner it would perhaps become more generally used ; its weight is very much against it when working with high powers ; still, for the advantages it affords it is an adjunct which is by no means to be despised.

BLANK EYEPIECE.—It will be found convenient, especially in examining the back of the objective to observe diffraction phenomena, cones of illumination, etc., that a blank or 'dummy' eyepiece be employed ; that is, an ordinary eyepiece mount having no lenses in it. The aperture in the cap must, however, be a very small one.

STANDARD GAUGES FOR EYEPIECES.

The following sizes were adopted by the Council of the Royal Microscopical Society on December 20, 1899, as the standard inside diameters of draw-tubes for microscopes, the tightness of the fit of the eyepiece being left to the discretion of the manufacturers ·

No. 1, 0·9173 inch = 23·300 millimetres.
No. 2, 1·04 inches = 26·416   ,,
No. 3, 1·27 inches = 32·258   ,,
No. 4, 1·41 inches = 35·814   ,,

No. 1 is what is known as the Continental size, which is made almost universally by Continental manufacturers, and has been supplied for many years. It has also been largely used by English manufacturers.

No. 2 is the mean of the sizes used by the English trade for students' and small microscopes.

No. 3 is the mean of the sizes used for medium-sized binoculars and for microscopes of a similar class.

No. 4. is the maximum size for long tube binoculars.

Two sizes only of these are in general use—viz., Nos. 1 and 3, and these appear to meet all requirements. Notwithstanding the choice given by the society, certain houses for reasons of

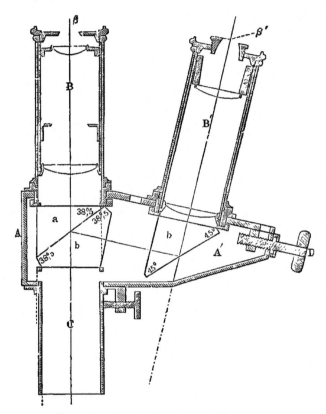

FIG. 32.—ZEISS'S BINOCULAR EYEPIECE.

their own make draw-tube fittings of diameters which do not conform to either of the above-mentioned standards, and those who may be contemplating the purchase of a microscope will be conferring a benefit on microscopy generally and directly on themselves if they insist that the microscope they select shall have one or other of the standard sizes of fittings for eyepieces, preferably No. 1 or No. 3, and refuse to accept any other diameter.

CHAPTER III

ILLUMINATION AND ILLUMINATING APPARATUS

Monochromatic Light and Light Filters.

ABSOLUTELY monochromatic light is a light of one refrangibility —that is, a colour of one uniform wave-length. As used in microscopy, monochromatic light means light with a small range of refrangibility, and it is important that its function should be clearly understood.

If white light is divided into its component parts by means of a prism or a spectroscope, a regular band of colours is produced, termed the spectrum, commencing with red at one end, followed by orange, yellow, green, blue, indigo, and finishing with violet.

In physical optics light is regarded as travelling in waves, the amplitude of each of which is very small, compared with the wave-length—not more than about 1 : 10,000. Now, the length of a light wave varies according to the portion of the spectrum that is used. At the extreme red end of the spectrum it measures $0^{.}76\,\mu$,* and the wave-length decreases through the range of colours until at the extreme violet end it measures $0^{.}39\,\mu$. From this it will be seen that nearly double the number of waves of light would be oscillating per millimetre with a violet light than with a red.

The numerical aperture of an objective is increased by the use of a dense medium enveloping the object and the front lens of the objective, as we have seen by the description of immersion objectives. The oil or other medium employed shortens the wave-length of the light used, whatever its colour, and when we

---

* $\mu = \frac{1}{1,000}$ of a millimetre, and is called a micron. There are $7\frac{1}{2}$ to 8 microns in the diameter of a human blood-disc.

use a light of shorter wave-length than it would have when passing through air, we increase the effective aperture of the objective. Accordingly the resolving power of a lens is increased by shortening the wave-length of the light admitted to it, and this is accomplished in either of two ways—(1) By employing blue instead of white light, or (2) by converting the lens into an immersion lens, and interposing a layer of oil instead of air between it and the object. For instance, if a microscopic objective were used with white light, and its limit of power to resolve fine structure were 50,133 lines per inch with such illumination, its limit would be 54,342 lines per inch with monochromatic blue light (line F).

A natural conclusion from these statements would be that the farther towards the violet the monochromatic light were used, the more marked would be the results obtained ; but although this is correct theoretically, it is not true practically. Microscope objectives are corrected for visual purposes for use with the brightest rays of white light, and if the extreme ends of the spectrum were employed—the objective not being calculated for these—rise would be given to spherical aberration, even in the best objectives, preventing the accomplishment of good work. If a lens were corrected for spherical aberration when used with light from the extreme blue end of the spectrum, under existing conditions of manufacture it would work at its best with the light for which it was designed, and if light lower down in the spectrum were employed, spherical aberration would be apparent. It must be borne in mind that light of extremely short wave-length is sensibly absorbed by glass, also the eyesight is not keen in extreme blue and violet lights, consequently the range of light that is practically available for monochromatic illumination is restricted.

Another advantage gained by the use of monochromatic light is, that as there is but one colour of the spectrum used, objectives of high-class make of the achromatic form are rendered practically equal to apochromatics by the removal of the secondary spectrum —that is, any chromatic aberration that may be present in the objective is annulled by the monochromatic light with which the illumination is effected. The more nearly the monochromatic light which is used approaches to that ray for which the spherical aberration in the objective is best corrected, the better will be the resulting definition. In some cases, so advantageous has this means of illumination proved, that, using two objectives of the same power alternately, one an expensive apochromatic, and the other an achromatic, it has been difficult to tell which was being employed.

True monochromatic light can be at the present time obtained by means of prisms only, and the most practical apparatus that is known to us is that designed by Dr. Spitta, and manufactured by Mr. C. Baker, of 244, High Holborn, London. It consists of a prism with a diffraction grating, slits, and condensing-lenses so arranged as to afford every facility for obtaining accurate effects.

Beautiful results with monochromatic light are to be secured by the use of a heliostat to reflect sunlight upon a slit and prisms.

Dr. G. Johnstone Stoney devised a most excellent heliostat actuated by means of a lever clock movement, and with fine adjustments by means of Hook's handles for setting, and a plane-parallel worked mirror. The brilliance of illumination by such means is very intense, and a spectrum 2 feet in length can be easily secured if the microscope be placed at some distance from the prisms. If the spectrum be allowed to fall on a sheet of white paper or cardboard at the position in which the microscope is placed, the exact tint of illumination that is required can be selected, and that alone utilized in the microscope. The use of a heliostat is necessarily limited in Great Britain, and especially so in London, on account of the few hours of bright sunlight that are available; but in countries where this restriction does not apply the heliostat is particularly to be recommended, not only for this especial purpose, but also for general microscopical work and in photography.

Many experiments have been made with a view to producing monochromatic light screens by means of pigments and the combination of coloured glasses, and thereby obviating the necessity for prisms. So far these attempts have been only partially successful, all that have been made passing light of more than one colour. The best results have been achieved by Messrs. Wratten and Wainwright, who supply for photomicrography especially a series of excellently corrected colour screens, or light filters, giving almost every desired colour effect. Very material assistance is afforded by these screens, not alone for actual monochromatic illumination, but also for general work, for they allow a large cone of illumination to be utilized without discomfort to the eyes. They are usually placed beneath the sub-stage condenser, and are employed in photo-micrography very considerably for neutralizing non-photographic colours in objects, and rendering the actinic and visual rays in an objective more nearly coincident ; also in visual work for minimizing light glare with large cones of illumination. This latter applies equally to apochromatic and achromatic objectives, the screen often producing stronger contrast and a crisper image than could be obtained without one.

The most advantageous colour for all-round visual purposes is the green-blue, and most effective light screens are made on a plan, originally suggested by Mr. J. W. Gifford, consisting of a film of gelatine stained with malachite green deposited on a circular disc of signal green glass, and having a protecting cover-glass for the gelatine fixed by means of a ring of cement. Fluid screens can also be made with great accuracy, and with care will last a considerable period, but it is particularly essential that they be protected from light, or they are liable to fade. The Gifford's screen, referred to above, in a fluid form is made by mixing a small proportion of malachite green in glycerine in a trough. The light from the illuminant that is to be used is examined spectroscopically through the medium in the trough, and the coloured fluid is added until the red end of the spectrum is absorbed ; if this be done exactly a minimum of loss of light occurs.

Another excellent screen is produced by making a saturated solution of acetate of copper, but a trough with an opening of at least $\frac{7}{9}$ inch back to front is necessary to obtain an effective colour with this fluid.

For photography, discs of light and dark yellow and various shades of blue glass are constantly employed, and particularly Dr. Spitta's 'pot' green glass disc, which is useful for all purposes, is an inexpensive and effective light filter. The great desideratum in a light screen is that it shall pass a large quantity of light. Mixtures could, probably, soon be made that would produce monochromatism, if only great transparency were not of importance.

Mr. Nelson states * that monochromatic blue light ' makes a difference of about 14 per cent. in the case of low apertures, but beyond those of 0·9 N.A. its influence in increasing resolution is so small as to be hardly worth taking into account. What it does effect is the sharpening and clearing of detail already resolved.'

This is no doubt because only a small part of the first diffraction image, seen when looking at the back of the objective with the eyepiece removed, is utilized by the lens, and that this part consists of blue light whether it be blue or white light that is employed for illuminating, so that in both cases there are the same materials for resolving the detail of the object, and the only difference is that there is a haze of light of lower refrangibility (or greater wave-length) also present, which forms a luminous veil over the whole field. It is this veil which is removed by using monochromatic light, and therefore the effect is to sharpen and clear the detail that is already resolved.

Sub-Stage Condensers.

A few years since all that was considered necessary in order to illuminate in a proper manner an object under examination was the mirror, perhaps in conjunction with the bull's-eye stand condenser ; and in many cases the mirror was hung on a tail-piece which could be moved in an arc round the centre of the stage, and by this means light at any angle could be reflected on the object. The day for this, however, has gone by, and anyone who requires to get even fair results must use a sub-stage condenser in some form or other. Especially does this apply to high-power objectives. Plenty of illumination can be obtained with the mirror only for low-power objectives, but beyond these the object becomes ill-defined and the field dark. More especially since the study of bacteriology has taken so prominent a position has the condenser come to the front. Without its aid it would be almost impossible to distinguish between different species of these minute organisms. To the leading members of the Royal

* *Journal of the Royal Microscopical Society*, 1893, p. 15.

Microscopical Society, and especially to the late Dr. Dallinger and Mr. E. M. Nelson, is due the steady improvement that has taken place in the optical qualities of the sub-stage condenser. The two gentlemen named were indefatigable in their appeals and demonstrations to microscopists, urging the pre-eminent position that it should occupy in manipulation, and the proper methods of using it.

One of the most largely used of condensers is a chromatic one, named the Abbe illuminator, originated by the firm of Carl Zeiss, and now supplied by nearly all opticians. It is made in two forms, one having a numerical aperture of 1·20, and the other of 1·40. The former is the more commonly employed, and is principally adopted by students. The optical portion is shown in Fig. 33. It gives a brilliant illumination, with the highest

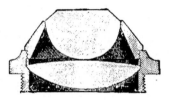

FIG. 33.—ABBE ILLUMINATOR. (1·20 N.A., OPTICAL PART.)

power objective, while with the lower powers, by removing the top lens, good results can be obtained. Its price is low, but it has the great disadvantage of not being achromatic, and having so large an amount of spherical aberration as to be almost useless for critical work, for although its total aperture is large, its aplanatic aperture is less than 0·50. Its popularity is to a great extent due to the facility with which it can be used, in consequence, probably, of its large field lens and its want of definite focus. Nevertheless, it fills a distinct position in microscopical work.

Modern critical research work necessitates the use of the best sub-stage condensers procurable ; consequently, the day for the chromatic condenser is passing, excepting for work of an unimportant nature, and English manufacturers have provided for every requirement in an ample manner. The importance of a well-corrected condenser can best be understood by the effect produced on the working aperture of the objective. To ascertain this the aplanatic aperture of the condenser and the numerical aperture of the objective should be added together, and divided by two, thus : If an objective of 1·30 N.A. is used with an Abbe illuminator having an aplanatic cone of 0·50, an effective working aperture is obtained of 0·90 (1·30 + 0·50÷2). If the same objective is used with a well-corrected condenser having an aplanatic aperture 0·90, an effective working aperture of 1·10 is obtained (1·30 + 0·90÷2).

Swift and Son make a range of condensers suitable for all powers, and the excellent Holoscopic series of condensers, computed by Mr. Conrady and manufactured by Watson and Sons, are well known.   Special mention should be made of the Uni-

Fig. 34.—C. Baker's Achromatic Condenser.

versal and the oil immersion of the latter series; these are constructed on the principle of the Holoscopic objectives, and have aplanatic apertures closely approaching their total apertures.   The Universal has a large back lens 0·92 inch diameter, an aplanatic aperture extending 0·90, and the equivalent focus of $\frac{4}{10}$ inch.   This power and aperture are the most generally useful that can be desired, and if the upper lens is removed by unscrewing, a condenser of longer focus is available for use with low-power objectives.   It is the best possible for medium powers ($\frac{1}{6}$ inch), and with oil immersion objectives is generally sufficient. It is only when oil immersion objectives of large numerical aperture are to yield their maximum effect that the oil immersion condensers are essential.

The achromatic condenser by R. and J. Beck, with the total aperture of 1·0, and an aplanatic aperture of 0·90, is an excellent one, as is also their immersion condenser.

Messrs. Powell and Lealand are entitled to special commendation for their early appreciation of the value of condensers having large aplanatic apertures, and they were many years in advance of other makers in the production of such condensers. They designed two of apochromatic form, one having a N.A. of 1·0 for use dry and the other an oil immersion having a N.A. of 1·40. Some of the condensers before referred to possess aplanatic apertures slightly in excess of those by Powell and Lealand, but they are beautifully corrected, and those who may be wishing to have condensers in keeping with their apochromatic objectives would find these admirable. Still, the value of a sub-stage con-

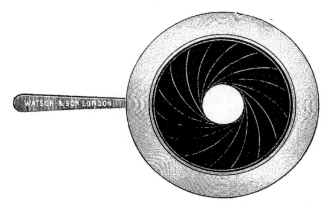

FIG. 35.—IRIS DIAPHRAGM AS FITTED TO THE CONDENSER CARRIER.

denser is not to be reckoned by its total numerical aperture, but by the solid cone that it will transmit, or, in other words, by its perfection of correction ; for its aplanatic cone alone can be employed for critical illumination.

If, therefore, an achromatic condenser gives a superior aplanatic cone to an apochromatic for practical work, it may generally be considered preferable, and, seeing that this is done, and that the achromatic form is very much less costly than the apochromatic, the microscopist may with assurance take the former, and be satisfied that he can perform with it all that possibly can be done.

In Fig. 34 the achromatic condenser that is shown is mounted on a carrier for the sub-stage. It is provided with an iris diaphragm similar to that illustrated in Fig. 35, by means of which any desired aperture may be quickly and exactly obtained. It will be found of utility to have the arc, through which the lever controlling the iris diaphragm travels, provided with a scale of divisions, so that results may be quickly reproduced or any special aperture may be obtained ; but for this purpose it is necessary that the diaphragm shall respond immediately on the pressure of the lever handle ; there must be no loss of time in the movement. The carrier is further provided with an arm, having a rotating cell, in which may be placed stops for producing dark-ground illumination in the same manner as with the spot lens, described on p. 107 ; also stops for obtaining oblique illumination for the resolution of the markings on diatomaceæ, and for holding tinted glasses or light screens. This form of carrier is applicable also to the Abbe illuminator.

A very efficient condenser can be oftentimes formed by fitting a low-power objective into a suitable carrier in the sub-stage. A small iris diaphragm, called the Davis's Shutter, having the universal male Society's screw at one end and a female screw at the other, together with a tube, fitting into the sub-stage and provided with the universal thread, into which the iris diaphragm may be fitted, is a very suitable carrier for the objective.

THE APLANATIC APERTURE of a sub-stage condenser is ascertained in the following manner :

The condenser is accurately centred, and both it and an objective are focussed in the usual way on an object mounted in *Canada balsam*, the edge of the lamp-flame being employed. We will presume that the objective has a numerical aperture of 0·5. The full aperture of the condenser is then used, and the object so placed that the balsam portion is still between the condenser and objective, but the object itself is not in the field. It would be well that a balsam-mounted object were always used for this experiment, as the result is slightly affected by different media. The eyepiece is now removed, and the back lens of the objective is examined. It may be found that it is completely filled with light, as in Fig. 36, under which circumstance the condenser has an aplanatic cone exceeding the N.A. of the objective. The aperture of the condenser can now be limited by means of a diaphragm, and an approximate value obtained for the size of diaphragm that is used. The edge of this diaphragm should be so set that its edge is just seen appearing at the margin of the objective, as in Fig. 37. The aperture of the condenser when used with this size of diaphragm is therefore a shade less than N.A. 0·5—say 0·45. An objective having a larger N.A., say 0·95, is now employed, and it will be found that the back lens of the objective is no longer filled with light. Theoretically, this is the condition under which the aplanatic aperture should be estimated, but when a flat flame of a lamp is presented edgewise, its image has corresponding depth, and when one part is focussed on the object, other parts of the image of the flame will necessarily be out of focus. There is therefore a certain range of adjustment of the condenser within which the effect (so far as it depends on focussing the light on the object) will

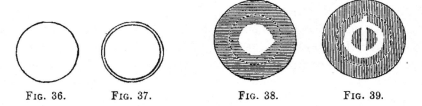

FIG. 36.   FIG. 37.   FIG. 38.   FIG. 39.

be pretty much the same. But these different positions give different apertures to the condenser as judged by the light reaching the back lens of the objective. The condenser should then be gently racked upward until the disc of light is at its largest (Fig. 38) ; until on a further movement of the condenser two black spots appear, one on either side of the middle of the disc (Fig. 39), which increase as the condenser is further racked up. *The last point before the appearance of the black spots furnishes the position in which the condenser has the largest aperture consistent with its outstanding spherical aberration not too much interfering with the highest results, and is the limit of the condenser for critical work.* Any further advance of the condenser gives merely annular illumination, which, of course, is to be avoided, excepting when stops are used.*

* E. M. Nelson, *English Mechanic*, November 16, 1888 (vol. xlviii., No. 1234), and ' The Microscope and its Revelations,' by Dr. Dallinger.

How to Use the Condenser.

The condenser requires almost as much care and skill in adjusting as the objective, for if it be improperly set up it will give rise to ' false images.' For an objective to work at its best the rule generally followed is to focus the image of the *edge* of the lamp-flame sharply upon the object on the stage, and this, with the modern condensers of large aperture, requires to be as accurately performed as the focussing of the objective upon the object; hence the value of the fine adjustment to the sub-stage (see p. 18). The following will be the procedure :

A proper microscope lamp, as described on p. 116, should be set in front of the instrument with the edge of the wick towards the microscope, and the light from the lamp may be allowed to fall directly upon the condenser, or a plane mirror may be used. The sub-stage condenser should now be centred, first having been placed in about the position that it will occupy when focussed. The centring cannot, of course, be properly effected without a centring sub-stage; but where there is only a fixed under-fitting, it is well to set the condenser at the position where it is most central. It is understood that the under-fitting is centred with the optical axis of the microscope when sent out by the makers; but owing to the fitting-tubes being more or less elliptical, it often happens, if the condenser is rotated in the under-fitting, that it will be central in one position only, and at this position it should be placed for working when there are no centring screws. Some condensers have fitted on top of them a removable cap with a very small pin-hole. This pin-hole should be focussed with, say, an inch or a ½-inch objective, and the condenser centred by it; the cap should then be removed from the condenser for working. Condensers not provided with this cap can, as a rule, be centred by using a diaphragm having a very small aperture at the back of the lenses, and focussing the aerial image of it with a ½-inch objective ; for this reason the iris diaphragm of the condenser carrier should be exactly axial with the condenser itself. This is necessary, too, for accurate working; but the easiest way of centring is to make a very small spot in the middle of the top of the lens with a pen and ink ; centre by this spot, and wipe it off. It will not make any difference to the performance of the condenser, and will insure accuracy and save time. Having centred the condenser, it should be racked up until it touches the under side of the slide, the objective being made to touch the cover-glass of the object on the upper side ; see that the diaphragm of the condenser is open, reflect the light with the mirror, and thus illuminate the field ; then rack the microscope body upwards until the object comes into view. If it is found that there is too bright a flood of light, the aperture of the condenser must be decreased a little by using a smaller diaphragm. Having focussed the object on the upper side with the objective,

FIG. 40.—IMAGE OF LAMP-FLAME.

it will be necessary to focus the condenser. Rack this downwards from the object very slightly until the image of the lamp-flame is seen in the centre of the field, the remainder being comparatively dark, as in Fig. 40. If now it be desired to have the whole field equally brilliant, a bull's-eye stand condenser may be interposed between the lamp and the mirror, the plane side of the bull's-eye being towards the lamp ; or the burner of the lamp may be turned round till the flat of the wick is towards the mirror. Where high powers are to be used, the object to be examined may with advantage be set upon the stage and focussed with the $\frac{1}{2}$-inch or other low-power objective, and the sub-stage condenser focussed upon the object. The high power may then replace the low power, and the condenser will be in adjustment. If it be found that the image of the lamp-flame is not in the middle portion of the field on exchanging the objectives, it will show that the objectives have not exactly the same centres, and the image must be set central with the high power by altering the position of the condenser by means of the centring screws of the sub-stage.

Later experience has shown that the method recommended by Mr. Conrady, and so ably demonstrated and advocated by Dr. Spitta, has distinct advantages. It does not regard so much the sharpness of the image of the lamp-flame in the field of view, but rather the carrying of the condenser beyond the focus to that point when the back lens of the objective is most fully flooded with light. In other words, the best illumination for critical work is secured at the stage immediately preceding the appearance of the black dots described in italics on p. 98. For this adjustment experience of the appearance of the back lens of the objective is essential, the eyepiece being removed for the purpose, and observation being made down the tube of the instrument. Acquaintance with the features of the back lens of the objective is a most desirable and really necessary acquisition for expert work.

Reliance may be placed on the first method described for accurate manipulation, and with fuller experience the second arrangement can be taken advantage of.

The next question is, What amount of light should be admitted from the condenser in order to see the object at its best? Mr. Nelson has suggested that the aperture of the condenser should be about three-quarters that of the objective, and in order to arrange this it will be necessary to remove the eyepiece from the microscope and look down the tube at the back of the object-glass, opening the diaphragm of the condenser to its fullest extent. Bearing in mind the size of the circle of light seen, gradually diminish the opening of the diaphragm of the condenser until one-quarter of the back lens of the objective is shut out; again put in the eyepiece, and the desired amount of illumination is arranged. The aperture employed should be varied slightly according to the transparency or opacity of the object under view.

When the condenser is centred and focussed, and the back lens of the objective is three-quarters filled with light, a *critical image* is obtained—that is, the objective is understood then to produce the finest results it is capable of.

Mr. Nelson's $\frac{3}{4}$-cone method of illumination has been almost universally accepted as a most practical one; but the following plan, which was suggested to the writer by a microscopical friend, has given very satisfactory results. On examining the back lens of the objective with a striated object, such as *Pleurosigma angulatum*, resolved and focussed upon the stage, there will be seen a central disc surrounded by six diffraction spectra similar to Fig. 41. With the $\frac{3}{4}$-cone illumination, the surrounding spectra will, in some cases, appear to overlap the central disc, and in others will not appear to touch it. Our plan is to open the diaphragm of the condenser only to such an extent that the spectra just touch, but do not overlap, the central disc. This would necessitate that in some instances we should employ rather less than a $\frac{3}{4}$-cone illumination, and in others rather more than a $\frac{3}{4}$-cone. We have not been able to observe that any loss of resolution results from this practice; but, on the other hand, in our opinion detail is more clearly seen, and appears crisper under these circumstances of illumination than any other. This system is especially advantageous when monochromatic light is used.

Fig. 41.—Back Lens of Objective.

As previously mentioned, it will be found of great advantage to become acquainted with the appearance of the back lens of the objective when working; many hints of importance may be gleaned from it, enabling manipulation to be effected with increased precision. For this purpose the 'blank' eyepiece referred to on p. 87 is a most useful adjunct.

When working with monochromatic light, the condenser must be focussed so that the whole of the light which is visible on the back lens of the objective when the eyepiece is removed shall appear as nearly as possible of the same colour.

Condensers having a numerical aperture of 1·0 and over require to be immersed in order that they may work at their full aperture—that is, a drop of immersion oil or Canada balsam must be placed between the top lens of the condenser and the object. It will be found generally that the condenser is a little too long in focus for continuity between the top lens of the condenser and the under side of the object to be maintained. Under such circumstances an additional thin 3 × 1 inch slip, or a piece of cover-glass, should be placed under the object, which will enable the oil contact to be maintained. The distance between the condenser and the object will vary according to the thickness of the slip on which the object is mounted, and the intermediate contact-glass will have to be selected accordingly. To use an oil immersion condenser effectively, the object must be mounted in some medium, and not dry upon the cover-glass.

As before mentioned, with the majority of condensers stops are supplied having the centres blocked out, as shown in Fig. 42, by means of which dark-ground and oblique illuminations are obtained. Dark-ground illumination gives a most beautiful effect to very transparent objects, such as infusoria, pond-life specimens, etc. In the form of carrier for condensers shown on p. 95, a cell is provided just above the iris diaphragm to carry the stops. One similar to *a* (Fig. 42) is placed in the cell, the iris diaphragm is opened completely, the condenser having been previously adjusted in the usual way, when it will be found that the object will be illuminated, but the ground on which it is seen will be black. Different objectives require stops of special sizes,

FIG. 42.—STOPS FOR CONDENSERS.

which may be readily made of blackened cardboard, cut to the most suitable size for working with the objective. An expanding iris stop, constructed in a similar manner to an iris diaphragm, by means of which a variable size of central black-patch stop is secured, is obtainable to fit most condensers, and will be found of great utility in black-ground illumination.

These stops can be further used for strengthening the contrast in the image with large cones of illumination and objectives having high apertures. This method does not minimize in any way the effective working of the objective, for with objectives of large aperture rays may be present which only impart brightness to the field, but do not contribute to making visible the fine detail upon the object. If less than half of the lateral spectra, as shown in Fig. 41, are seen on looking down the tube at the back lens of the object-glass with a striated object in focus, then the central portion of the direct beam or central disc has no lateral image corresponding to it in the portions of the spectra that are visible. Under these circumstances that central portion of the central disc in no degree contributes in enabling the detail to be seen, but only produces a haze ; by blocking it out the haze is removed, and there is a great improvement in the resulting definition. This produces oblique illumination in all azimuths.

THE CHOICE OF A CONDENSER.—That the condenser is an absolute necessity cannot be too strongly impressed. No good results can be obtained without it.

Condensers, like objectives, not only vary in aperture, but also in power, and the higher the power of the condenser the smaller will be the image of the lamp-flame that it transmits. Consequently, if a condenser of high power is used with a low-power objective, the illuminated portion of the field will be exceedingly small, while if a low-power condenser is used with a high-power objective, the image of the lamp-flame is so magnified that the whole field is bright, and it is not easy to tell when the condenser is exactly focussed. Furthermore, under such circumstances as the latter, it is impossible to get the best effect with the objective. It has usually been recommended that two condensers, one of high and the other of low power, should be included in a complete equipment, but the new types of condensers previously referred to cover such a large amount of ground that the average microscopist really requires only one. Choose a condenser that gives an aplanatic cone of 0·90 N.A., and if the major portion of the work is to be with objectives of low and medium magnifications, one of low power should be selected ; if principally with high powers, a corresponding high-power condenser will be necessary. It must be remembered that by removing the uppermost lens by unscrewing its cell, the remaining combinations of a high-power condenser form a very serviceable low-power condenser.

Oil immersion condensers are of especial utility where objectives of the largest aperture are employed, and these, again, in several instances work well as dry condensers, and with the top lens removed become low-power condensers of moderate aperture.

If the advice given be followed and the condenser used intelligently, reliance will be placed in the work performed, and far superior results secured than would be possible with a badly corrected chromatic condenser.

Dark-Ground Illumination.

Reference has been already made to the Ultra-Microscope, and under this heading it will be necessary to describe the recently introduced contrivances for the obtaining of dark-ground illumination with high-power objectives, and effected in such a way as to render visible subjects which could not be observed by other means. The effect is produced with all the systems that are employed, of illuminating the object with rays more oblique than any the objective can receive—that is, no direct light enters the microscope, and the field is therefore dark. The diagram (Fig. 43) will give a general idea of the manner in which this takes place. Although the oblique light rays do not enter the objective, they cross the field in which the objects lie, and, striking the objects, the light is reflected, refracted, and diffracted. This light entering the objective causes the objects to stand out brightly on a dark background in the same way as when a ray of sunlight enters a dark room and renders visible the extremely minute dust particles. Thus objects of excessive minuteness which would be invisible by ordinary methods of illumination are clearly perceived.

These illuminators have been found of great value in the examination especially of bacteria in the living state, enabling identification to be at once made, and rendering unnecessary the tedious preparation of preparing, staining, etc., which would be involved for ordinary observation by the usual methods. In consequence of the similarity of results between this and the ultra-microscope, the dark-ground illuminators have been called by the same name; this, however, is inaccurate, and the dark-ground illuminator must be regarded separately and distinctly. Exact instruction for the use of these illuminators is given by the various makers, and they should be studiously followed; but there are general principles applicable to all which may with advantage be indicated.

The illuminating rays have the numerical aperture extending from 1·0 to 1·45. The objectives employed must, therefore, have a numerical aperture less than 1·0, otherwise they could receive the direct rays, and dark-ground illumination would no longer be obtained. Obviously immersion oil must be placed between the illuminator and the object-slip, or the aperture in excess of 1·0, which is the maximum of air, will not be obtainable. It follows also that the object itself must not be dry-mounted, but contained in water or oil, which it is important should be perfectly clean and clear. This layer of medium, as well as the object itself, must be as thin as possible ; further, the object-slip must be about the thickness prescribed for each illuminator, and must be free from scratches and other defects. The illuminator must be accurately centred to the objective, the method given on p. 99, by placing a spot of ink on the top surface of the illuminator for the purpose, being probably the simplest one. The more nearly the objective approaches to 1·0 N.A., the more fully are the best effects realized.

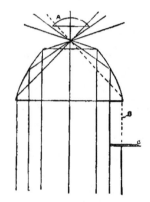

FIG. 43.—DIAGRAM SHOWING RAYS PASSING THROUGH IMMERSION PARABOLOID.

A, Front lens of objective ; B, rays of lowest numerical aperture.

If it is desired to use the higher magnifications given with oil immersion objectives, such objectives must have an appropriate stop inserted from the back to reduce the numerical aperture to 1·0. The effect is enhanced by the use of a brilliant source of light and a well-corrected bull's-eye; but even an oil lamp with a bull's-eye will enable effects to be obtained and work to be done.

The illuminators of paraboloidal construction as introduced by Carl Zeiss, and made in a modified form by Watson and Sons and R. and J. Beck, possess great advantage over ordinary refracting condensers and dark-ground illuminators of more complicated construction, inasmuch as the iris diaphragm cuts out the rays of low numerical aperture first, for as the diagram (Fig. 43) shows, the paraboloid may be said to turn the incident bundle of light inside out. In other words, the rays passing through the outermost zone of the iris diaphragm are those of *lowest numerical aperture*, whilst the most oblique and most valuable rays are those entering near the centre of the paraboloid. The result is that the amount of light can be reduced without losing those invaluable rays of maximum obliquity. These illuminators are not suited for objectives which have a lower numerical aperture than 0·48.

It may be asked why equally effective results cannot be obtained with the ordinary sub-stage condensers and dark-ground stops. The paraboloid gives a *blacker* background than is possible with a condenser, because the numerous polished surfaces of the latter reflect light upon each, a proportion of which finds its way into the central part of the cone of rays produced by the condenser, and impairs the blackness of the field. The immersion paraboloid must not be confounded with the form of paraboloid used many years ago for dark-ground illumination.

THE SPOT LENS.—Before the sub-stage condenser came into general use the spot lens was largely utilized for obtaining dark-ground illumination. It has, however, been to a considerable extent superseded, owing to the perfection in which the same effect can now be obtained with the condenser. It is intended for low powers only up to ½ inch.

For one class of work the spot lens is especially advantageous. Most sub-stage condensers have a very short focus, and if organisms in water in a trough are being examined, it is impossible to focus the condenser accurately through the trough and its contents. A spot lens has a longer focus, and gives under these circumstances the best results. With it a plane mirror and the flat of the wick of the lamp should be used ; the sub-stage that carries it should be moved up and down until a perfectly black ground is obtained. If additional brilliancy is required on the object, a stand condenser interposed between the lamp and the mirror, with the convex side of the condenser towards the mirror, will give a brighter effect.

THE INDIAN INK METHOD.—There is a recently discovered method of showing living bacteria with a black background which anyone may make use of without extra apparatus beyond a supply of suitable Indian ink. Burri first used this method in order to obtain an absolutely pure culture, as the bacilli are ' stained ' without being killed, and the growth of a single bacillus can be observed under the microscope. Bacilli and leucocytes, as well as spirochætes, show up as clear spaces. The following is the technique recommended by Frühwald, of Munich :

The surface of the suspected lesion is shaved off with a scalpel until a drop of serum is obtained which is not too darkly coloured with blood. A loop of this is mixed, upon a microscopic slide, with a drop of commercial Indian ink. The mixture is then spread with the edge of a coverslip. The film dries within the minute, and it can be examined immediately with the oil immersion lens. The spirochætes are seen as bright spirals on a dark brown field. Leucocytes, bacteria, and other spirochætes are also seen as clear spaces. The effect produced is dependent on the fact that the particles of Indian ink are smaller than the bacteria, and hence, when a mixture of the two is allowed to settle on a slide, the bacteria remain as clear transparent spaces on a dark ground.

The Polariscope.*

This consists of two parts, each composed of a Nicol prism of Iceland spar in a suitable mounting—one called the polarizer,

FIG. 44.—POLARIZER.   FIG. 45.—ANALYZER.

which fits into the sub-stage, and the other the analyzer, which is usually inserted between the nosepiece of the microscope and the objective. By its means light is split up into its component parts, and most beautiful colour effects are obtained. The polarizer has a flange beneath, by which it can be rotated, and in this way the colours are varied. In examining certain chemical crystals, geological slides, etc., it brings into view structure which without it would hardly be detected, and for this it is largely

* Further instructive information regarding the polariscope is given in the treatment of Petrological Microscopes on p. 219.

used in analytical work. In some instruments the analyzer prism is fitted in the body. This is rather an inconvenience, unless the instrument be designed especially for petrology. For a binocular microscope, however, if it is placed between the nosepiece and the objective, it causes a separation between these two, which interferes with the performance of the binocular prism, because the closer the back lens of the objective can be brought to the binocular prism, the more perfect will the vision be. Under these circumstances the monocular tube only is generally used ; or the analyzer prism can be mounted over the top of the eyepiece of the monocular tube. The theory of polarization is fully explained in connection with Petrological

FIG. 46.—DARKER'S SELENITES.

Microscopes on p. 219. For use with the polariscope, varieties of tints and a background of colour can be obtained by the employment of selenite films. These, in the cheapest form, are mounted in the same way as ordinary microscopic objects ; but a still greater variety of effect can be obtained by having selenites fitting into a carrier to come between the polarizer and the stage in a sub-stage microscope. We illustrate one (Fig. 46) by R. and J. Beck. In this form each of the selenites is provided with a ring which rotates. The three being one over the other, either two or all three can be rotated together or in opposite directions to one another, and the effect is most striking. An inexpensive modification of this is made by Swift and Sons and others, called the mica-selenite stage, as shown in Fig. 47. This consists of a film of mica made to rotate in a brass plate, upon which the object is laid, and beneath it is a carrier with three separate selenites. These can each be pushed separately beneath the mica and the latter rotated. By this means all the different tints obtainable with any number of selenite films can be produced. It can be employed on any

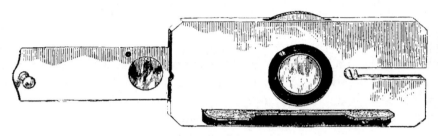

FIG. 47.—MICA-SELENITE STAGE.

microscope. To get greater brilliancy the polarizer can be made to fit into the sub-stage condenser on the under side, and the Abbe chromatic and achromatic condensers referred to previously are particularly suitable for this arrangement.

The Bull's-Eye Condenser.

Many objects, being opaque, cannot be viewed by light from beneath, and consequently have to be illuminated from above. In order to do this a bull's-eye condenser is necessary. This usually consists of a plano-convex lens mounted on a stand, as shown in Fig. 48. This has a ball and socket joint and a sliding telescopic tube, by means of which the lens can be placed in any desired position. Both Mr. Nelson and Mr. Conrady have introduced improvements in the construction of bull's-eye condensers in order to reduce the large amount of spherical aberration which is a necessary accompaniment of the single lens. These improved forms consist of either two or three lenses in combination, and the advantage obtained, especially in photography, is well worth the additional outlay.

THE USE OF THE BULL'S-EYE CONDENSER.—The diagram on p. 112 will show that effects are produced by the bull's-eye according to the position the light is placed in relation to it. For the illumination of opaque objects on the stage of the microscope, a brilliant point of light is required. Fig. 49, B, shows that to secure this the bull's-eye should be set midway between the illuminant and the object. It will be obvious that the light and bull's-eye must be in alignment with the object, and it will be generally found more convenient if the light be set somewhat above the level of the stage, so that the shadows produced by the

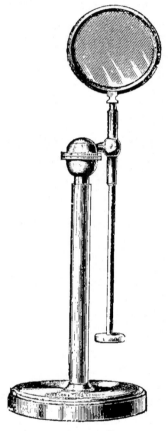

Fig. 48.—STAND CONDENSER.

object may not be too pronounced. For certain work, however, it is necessary to obtain a parallel beam with a bull's-eye ; to do this the light must be set in the principal focus of the bull's-eye, as shown in Fig. 49, C. This arrangement will often be found useful for illuminating objects viewed with low powers, from beneath the stage, when no sub-stage condenser is employed ; and for low-power photography this is particularly convenient.

If now the light be placed some distance from the bull's-eye,

as in Fig. 49, A, the focal point will be seen to be shorter. The
effective working of the bull's-eye can only be obtained by knowledge of these rules and practice. It will be found advantageous
to conduct some experiments, holding a piece of card in the path
of the rays, and observing the effects.

It must be clearly stated that no matter what lens is used, no

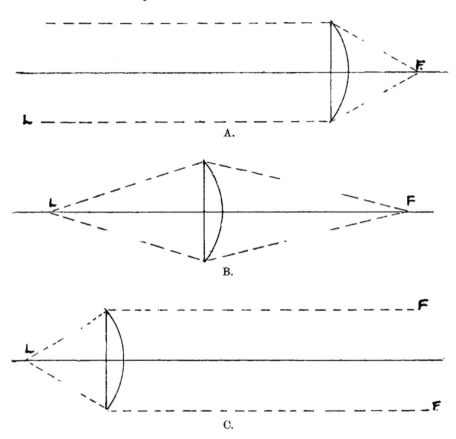

Fig. 49.—Diagrams showing the Effect of placing the Bull's-Eye Condenser at Different Distances from the Source of Light.

increase of light is obtained in excess of that of the illuminant;
all that it does is to collect and concentrate that light ; and
Mr. A. E. Conrady, in connection with his Watson-Conrady
achromatized bull's-eye, has pointed out that, although not of
large diameter (2¼ inches), it utilizes all the rays from the illuminant which it is possible to employ for microscopical purposes.

The bull's-eye condenser is often used in conjunction with the sub-stage condenser, to enable the field to be evenly illuminated with the latter, as mentioned on p. 100.

The Parabolic, or Side Silver Reflector.

To enhance the effects obtained with opaque objects with stand condensers, the side silver reflector is most advantageous. The arm on which this is mounted is attached to either the stage or limb of the microscope, or fitted between the nosepiece and the objective. The reflector consists of a highly-polished silver parabolic speculum. This reflector is placed by the side of the object, and light is directed from the lamp through the bull's-eye on to its centre, and then thrown by the reflector on to the object. Most brilliant opaque illumination may be obtained by this means.

The Vertical Illuminator.

The merits of this piece of apparatus were but little appreciated until the study of metal surfaces microscopically was taken in hand. Its use had been almost exclusively restricted to ascertaining whether the specimens were mounted in contact with the under surface of the cover-glass—information which is of importance when an oil immersion object is being employed on an object mounted dry, for if the specimen is not adherent to the cover-glass it cannot be seen at all with this illuminator. It is also used for the resolution of the markings on diatoms that are mounted dry on the cover-glass ; in this case it is, however, only of value with immersion objectives.

In metallurgical work it has rendered permissible the examination of surfaces of metals with high-power objectives. The surface to be examined is highly polished and then etched with acid, liquorice juice, or other medium, as will be found described on p. 196, and no covering-glass at all is used. It is made in two patterns, one with a prism, and the other with a disc of cover-glass. Metallurgists usually have the two forms, finding one more serviceable than the other under certain conditions and with different specimens. They both have their distinct value.

Generally speaking, for this class of work the prism pattern is the better for low-power objectives, and the disc form for the high powers. The latter is illustrated in Fig. 50, and may be either built in the body of the microscope or screwed to the nosepiece of the microscope above the objective ; the one illustrated is arranged for the latter. The disc of cover-glass attached to a little clip having a milled head, *a*, is placed through a slot in the fitting, *b*, and set at an angle of 45° to the optic axis. Light is received on this disc of glass through the small opening in the body of the fitting, *b*, and it is totally reflected through the object-glass on the object, the objective acting as its own achromatic condenser. In order that the light may be focussed on the object, the lamp-wick from which the light is being obtained must be the same distance from the reflector as the latter is from the diaphragm of the eyepiece, if a positive eyepiece is being

b

FIG. 50.—VERTICAL OR DISC ILLUMINATOR.

used, or to the eye-lens, if a Huyghenian or negative eyepiece is employed.

To save a frequent readjustment, some workers have the lamp fitted with a projecting tube, which, when in contact with the side of the microscope, gives the exact distance at which it should be set in relation to the vertical illuminator.

To know what can be done with the vertical illuminator, it is necessary that a fitting be attached to it into which stops and diaphragms can be placed, and for the effective working a bull's-eye condenser is set in such relation to the lamp that the aerial image of the lamp is produced at a distance from the vertical illuminator equal to that from the vertical illuminator to the diaphragm or eye-lens of the eyepiece, according to the eyepiece that is in use. The position of the aerial image must be ascertained by the use of a white card screen ; by this means the same effect will be produced as when a stop or diaphragm is placed immediately over the vertical illuminator.

GLARE WITH THE VERTICAL ILLUMINATOR.—Metallurgists are frequently troubled with glare which is constantly present in the field of view when using a vertical illuminator, and it is often so obtrusive as to seriously diminish sharpness of definition and perception of detail. This is obviated to a great extent by having the objectives set in special mounting, so that the back lens approaches as nearly as possible to the reflector of the vertical illuminator; but it may be modified—in fact, for practical purposes, eliminated—by using an iris diaphragm quite close to the source of light or in front of the bull's-eye, and reducing the aperture through which the light passes to such a point as will remove the objectionable glare. This will not affect the brilliance of the actual image, for only the precise amount of light which can be utilized, no less and no more, will pass through the objective.

CHAPTER IV

ACCESSORY APPARATUS

The Lamp.

IT is most important that the microscopist should possess a good and suitable lamp, otherwise he cannot work to the greatest advantage. The amateur will often be found working with a reading-lamp or an ordinary oil-lamp, but good work can never be done conveniently by this means. There are two or three important points which must be borne in mind. In the first place, if light is proceeding from the one illuminating point only, and the remainder of the room is dark while using the microscope, a great deal better effect can be produced than if the whole room be illuminated. In the next place, a small brilliant source of light is far better than a large one. In recent years special attention has been paid to this matter, with the result that lamps have been constructed with which the best work may be accomplished. The following are desirable features which should be embraced by a good microscope lamp: The reservoir for oil should be large in diameter and flat, so that the light may be brought down very close to the table. For this reservoir glass is usually preferable to metal, it being much cleaner, and the worker is able to tell when his oil is getting exhausted; whereas with a metal reservoir, unless careful reckoning is kept, in the middle of some important observation the light may go out from want of oil. A $\frac{1}{2}$-inch wick is generally found to be sufficient. We strongly deprecate the use of glass chimneys. They are always liable to get broken very easily, and become a source of expense, in addition to which, if away from town, there is a possibility of not being able to get the right kind, and so work may be delayed. Far better will be found the

116

metal chimneys now made by nearly all opticians, with a carrier
for a 3 × 1 or 3 × 1½ inch slip. It is obvious that if the slip be
broken it can be immediately replaced, it being part of the
microscopist's average stock. It is also desirable that the bar
on which the lamp is raised and lowered on the stand should be
a square one. If round
in shape, the lamp is apt
to swing round on the
stand and the whole to
topple over. This is an
impossibility with the
square bar. Such a lamp
is shown in Fig. 51, by
Swift and Son, and modi-
fications of it can be obtained from most dealers.

A similar model of
lamp which has rack-
work and sliding adjust-
ments in the vertical
direction, and a quick
acting screw for lateral
adjustments, is also ob-
tainable. Its mechanical
arrangements enable the
light from the lamp to
be shifted just the slight
amount that is often
needed, and obviate the

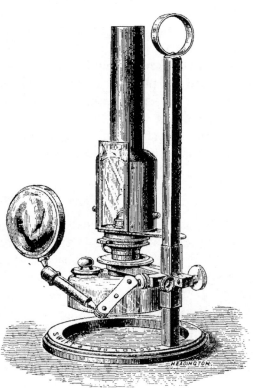

FIG. 51.—MICROSCOPE LAMP WITH METAL
CHIMNEY.

necessity for moving the lamp itself, as would otherwise have to
be done.

ELECTRIC LAMPS.—The increasing use of electric light has
caused some attention to be given to this form of illuminant
for microscopical work. There are several very handy patterns
that can be connected to the ordinary house circuit. It is not
suited for critical work, but the average observer has so much
to do for which such a lamp is suitable and immediately available
that it is becoming largely used. The lamp should have a frosted
bulb, and the stand should permit of universal movements.

The Nernst electric lamp has been found a brilliant and useful source of light. The narrow filament is a cause of inconvenience but its utility is such that it is the subject of constant improvements and experiment. The opticians' catalogues should be referred to for latest developments.

INCANDESCENT GAS LAMPS. — A very brilliant and efficient illumination can be obtained from an incandescent burner on a table-stand of suitable height properly shaded. The reticulations of the mantle would render it objectionable, but this can be obviated by placing an iris diaphragm a short distance from the light in the same manner as is described for the lamp above, or a narrow slit would answer the purpose equally as well; the diaphragm or slit would then be focussed by the sub-stage condenser, and treated as the source of light.

Objective Changers.

THE REVOLVING NOSEPIECE. — Time-saving arrangements will be found valuable in working, and the nosepiece is one of these;

FIG. 52.—THE DUSTPROOF NOSEPIECE FOR THREE OBJECTIVES.

it is, in fact, almost a necessity, especially where constant change from a low- to a high-power objective is necessary. It is screwed into the fixed nosepiece of the microscope, and is made to carry either two, three, or four objectives, termed respectively the double, triple, or quadruple nosepiece. Each of the objectives can in turn be rotated into the optical axis, thus saving the necessity of unscrewing an objective and screwing another on in order to get a variation of power. In hospitals, laboratories, etc., it is usual to have one of these fitted to every instrument. The dustproof pattern has found special favour, rendering unnecessary the removal of objectives after use. One of these is illustrated (Fig. 52). They are to be had made of an aluminium alloy which is extremely light, reducing the strain on the body-tube. Any microscope having the universal size of thread for objectives will carry a revolving nosepiece; no special adaptation is required. When the revolving nosepiece is screwed home, the objectives not in use must point towards the middle of the front of the stage, otherwise in rotating the objectives they are, with low powers, apt to foul the rackwork bar of the microscope.

OBJECTIVE CHANGERS BY ZEISS.—It will be readily perceived that however accurately a revolving nosepiece may be made, the same portion of the object will not occupy the centre of the field when different objectives are brought into use if the objectives themselves have not identical centres. Variations in this respect frequently occur, especially with objectives by different

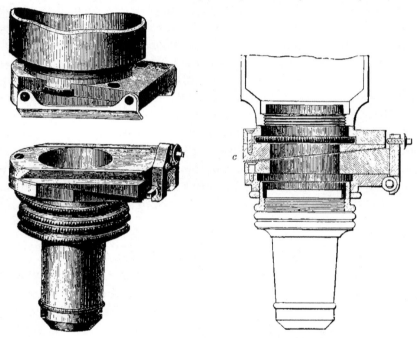

FIG. 53.—SLIDING OBJECTIVE CHANGER, BY ZEISS.

a, Tube-slide ; *b,* objective slide with objective attached ; *c,* in section, tube slide and objective slide united.

makers, and recognizing this, Carl Zeiss introduced an objective changer consisting of two parts, one called the ' tube slide ' (*a*), which is screwed into the nosepiece end of the microscope-tube, and the other the ' objective slide ' (*b*), to which the objective is screwed. A separate objective slide is required for each objective.

The accompanying illustrations give an excellent idea of the construction. It will be seen that the two ' slides ' are united by a sliding fitting, the upper of which is inclined downwards, so that when an objective is withdrawn it is lifted from the object. Attached to the lower slide are two screws, one of which serves to centre the object longitudinally, and the other transversely. By this means all objectives can be set to a common centre and permanently retained so. The simplest mode of procedure is to set a small specimen in the centre of the field with the objective of highest power, without either of the 'slides' in position—that is, the objective should be screwed into the microscope direct. This is done so that the correct centre may be first ascertained. Then remove the objective and fix the 'tube slide' to the microscope, and screw the same objective to the 'objective slide,' and if, when in position, the object no longer occupies the exact centre of the field, the set screws are used until it does so. The next highest objective should be treated in the same way, and so on with other objectives. By retaining each objective in its own 'slide,' accuracy is maintained. This device is particularly convenient with microscopes having a concentric rotating stage, the object always remaining in the field. This is a most excellent objective changer, but its use demands care and more leisure than the laboratory worker can usually bestow; hence the revolving nosepiece is far more generally used.

The Nosepiece Iris Diaphragm, or Davis's Shutter.

This is a very compactly made iris diaphragm, which is placed between the nosepiece of the microscope and the objective. Its

FIG. 54.—DAVIS'S SHUTTER.

special function is to enable the aperture of an objective to be decreased, so that it may be used with dark-ground illumination, or to increase penetration when examining objects having several planes. For photographing opaque objects with low powers it enables the appearance of a small round object, such as a moth's egg, to be taken quite sharply. A similar result has been attained by mounting an iris diaphragm between the lens combinations of low-power objectives.

The Davis's shutter is furthermore very useful for examining and experimenting with the diffraction spectra seen on looking down the microscope-tube at the back of the objective, when the eyepiece is removed and a striated object is being examined.

This iris diaphragm must have its aperture perfectly central and the threads quite true. The aperture of the iris, when completely opened, should be as great as it is possible for the inside of the mount of an objective to be, but the box of the iris diaphragm must not be so large as to touch the bearings in which the tube of the microscope is raised and lowered. It is well to have the lever of the iris diaphragm working in front of the microscope-tube, and so that this may be easily arranged the body of the iris should be so mounted that it may be rotated, but should be very stiff in its rotary movement.

Camera Lucida.

This is designed to assist in drawing objects seen in the microscope. Photo-micrography has to a large extent super-seded it, but does not depict the several planes of an object in the realistic manner that is possible in a drawing; still, there are a great many who prefer this method to any other. Dr. Beale's neutral-tint reflector, which is supplied by all the opticians, is the cheapest, and a very good form. It consists essentially of a neutral-tint glass—in which the image of the object is reflected —mounted in a frame to fit over the eyepiece. The method of using it is as follows: The microscope is set in a horizontal position, with the centre of the eyepiece 10 inches from the table. Illumination is arranged in the ordinary way. The cap of the eyepiece is removed, and the neutral-tint reflector is fitted in place of it, and is so arranged that the centres of the neutral-tint glass and the eye-lens of the eyepiece are in alignment, the former being set at an angle of 45°. On looking on this neutral-tint glass from the upper side, a disc of bright light will be seen on it, and if a piece of white paper be spread below on the table, on further examination the outlines of the object will appear to be upon the paper. If a pencil be now taken, the specimen can be sketched in its magnified form. This will be found somewhat difficult at first, nearly every worker seeming to find it necessary to work in some special manner of his own; but the secret of success is to arrange the balance of illumination by turning the lamp-wick up and down until a degree of light is found at which the pencil-point and image can be distinctly seen. When using students' microscopes, the eyepieces of which have no caps, it is usual to remove the eyepiece, fit the tube of the reflector to the outside of the top of the draw-tube, then reinsert the eyepiece and set the neutral-tint glass in position. The tinted glass is usually mounted on an arm which has a joint, so that it may be turned out of the way when not required without detaching the piece of apparatus from the microscope. The distance of 10 inches

FIG. 55. ABBE CAMERA LUCIDA.

between the eyepiece and the table is maintained, whether the microscope has a 6 or a 10 inch tube-length.

The Beale's neutral-tint possesses the disadvantage of reversing the image that is seen with it. Mr. Ashe, therefore, devised a modification known by his name, which overcomes this defect, while maintaining the simple principle of the Beale's pattern. It was described in the *Journal of the Quekett Club*. Of more expensive description, but considered the best at present made, is the Abbe camera lucida (Fig. 55). The microscope may be used in a vertical or any inclined position with this apparatus. Its construction and the manner of using is as follows : Mounted in a cap, which is fitted immediately above the eyepiece, are two right-angle prisms ; these are cemented together and form a cube. One of the cemented surfaces is silvered, but a small central disc is left clear, through which the object is viewed in the ordinary way. The prisms are so set that the image of the paper on which the drawing is to be made, and which is reflected by a mirror to the prisms, is by them conveyed to the eye. Thus the pencils of light reach the eye coincidently from both the microscope and the paper, and when drawing the object the pencil-point appears in the field of view very distinctly, and the minutest details can be exactly traced. Low-power eyepieces should be used with this camera lucida. There are many other forms peculiar to individual makers, possessing more or less merit, some of which may be used with the tube in any position, the Swift-Ives pattern, made by Swift and Son, being particularly efficient; it can be used with the microscope set vertically or inclined.

The Measurement of Objects.

There are three ways in which this may be effected ·

1. By having the stage divided—applicable to mechanical stages only.
2. By means of a camera lucida and a stage micrometer.
3. By means of eyepiece and stage micrometers.

1. If the movements of a mechanical stage are divided and read by verniers to very small parts of an inch or millimetre, the measurement of an object can be effected by having in the eyepiece a disc of glass with a diamond-cut line across the centre. The object that it is desired to measure is set with one point exactly against the diamond-cut line, which, of course, will appear in the field, and the readings of the stage divisions taken. The stage is then slowly moved along by means of the milled head until the other edge of the specimen to be measured is exactly touching the line. The readings of the stage divisions are again taken, and by subtracting one from the other the measurement will be ascertained. For quick work, and without extraneous appliances, this is fairly accurate, and largely used.

2. *Camera Lucida and Stage Micrometer.*—A stage micrometer usually consists of a number of lines photographed, or ruled with a diamond on a slip of glass to the scale of $\frac{1}{100}$ or $\frac{1}{1000}$ part of an inch, or the $\frac{1}{10}$ and $\frac{1}{100}$ of a millimetre. This is put on the stage and focussed like an ordinary object. The camera lucida is then fixed to the eyepiece, and the micrometer lines

are projected on to a piece of paper in the same way as when drawing an object explained on p. 122. The lines so projected are then measured, and supposing the lines of the micrometer, which are $\frac{1}{100}$ of an inch apart, appear when drawn on the paper 1 inch apart, it is at once known that the magnifying power in use is 100 diameters. The object may be measured in the same manner. Measurements should be taken about the centre of the field, and not towards the edge, especially with high powers, as, owing to curvature of the field, the outer edges appear more highly magnified than the centre.

3. *The Eyepiece Micrometer and Stage Micrometer.* — The stage micrometer, as previously described, is placed on the stage, and a somewhat similar micrometer is put into the eyepiece. This latter is generally divided into hundredths of an inch, but no exact value is needful so long as the lines are equidistant. On focussing the stage micrometer the two sets of lines will appear in the field at once. It is now desirable to ascertain how many divisions of the eyepiece micrometer are included between one of the spaces—that is, $\frac{1}{100}$ of an inch—of the stage micrometer. Perhaps it will be found that there will be several lines of the eyepiece micrometer and a fraction in that space, and in order that this fraction may be obviated the draw-tube should be slightly pulled out, which will give, of course, an increased amplification, until a certain number of the lines on the eyepiece micrometer are exactly equal to a division or divisions on the stage micrometer. We will imagine that the number of eyepiece micrometer lines that fill $\frac{1}{100}$ of an inch of the stage micrometer is five. The stage micrometer is now removed, and the object to be measured replaces it. The lines of the eyepiece micrometer will still be seen in the field, and bearing in mind that five of these lines equal $\frac{1}{100}$ of an inch, any part of the object can at once be measured. It must be remembered, however, that with every objective and at every tube-length an estimation of the value of the eyepiece micrometer is necessary.

To give greater facility and accuracy, a form of eyepiece micrometer is used, devised by Jackson, which is fitted in a frame, and by means of a micrometer screw traverses the object. If there be no mechanical stage to the instrument, it is very difficult to set a special part of the object against the micrometer

for measurement, especially with high powers. This form of
micrometer surmounts this difficulty. The ordinary eyepiece
micrometers necessitate no alteration to ordinary eyepieces,
because they rest on the diaphragm inside, but the Jackson form
requires that the outer tube of the eyepiece shall be cut to receive
the carrier for the micrometer. Fig. 56 shows an eyepiece with
the Jackson micrometer, *m*, in position.

There is yet another form of eyepiece micrometer, called the
Ramsden screw micrometer, which consists of an eyepiece
containing two webs or wires, one fixed, the other travelling by
means of a screw having 100 threads to the inch. The milled
head of this screw is divided into 100 parts. Across the field are

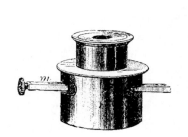

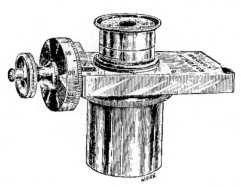

FIG. 56.—JACKSON MICROMETER
FITTED TO EYEPIECE.

FIG. 57.—RAMSDEN SCREW
MICROMETER.

very small equidistant V-shaped teeth, the interval between each
of which corresponds to one complete revolution of the milled
head. The value of these teeth is taken against the stage
micrometer, and the object placed on the stage. One edge of the
object is then brought against the fixed wire, and the travelling
wire moved to the other part that it is desired to gauge. By
then counting the number of intervening teeth, and reading the
fraction on the milled head, it can at once be ascertained what
magnifying power is used. This is considered the most
accurate and precise method of working, but it is an expensive
piece of apparatus, and with care one of the previous methods
named will be, as a rule, sufficient.

Persons having abnormal vision are likely to make errors in
measuring. To obviate this, a cap carrying a lens that will

correct the abnormality should be placed over the top of the eye-piece, as described on p. 84, while measurements are being taken, but usually in eyepieces arranged to carry micrometers, and in the Ramsden screw pattern, an adjustment is provided for sliding the eye-lens to and from the micrometer scale, so that it may be sharply focussed.

Troughs, Live-Cages, Stage Forceps, etc.

Troughs.—These are made of various materials, including glass, vulcanite, brass, etc., and are used in the examination of infusoria and animalculæ alive under the microscope. The essentials of a trough are that a medium power, say $\frac{1}{2}$ inch at least, can be used, that it may be easily cleaned, and that if broken it can be repaired. The ordinary commercial glass troughs unfortunately do not meet these requirements. They are difficult to clean, they

FIG. 58.—BOTTERILL'S TROUGH.

are invariably hard to mend when broken, and they very often leak when water is put in. The one that we have found most serviceable is the Botterill's trough, as shown in Fig. 58, which consists of two vulcanite plates between which are placed slips of glass, which are separated by an india-rubber band, small bolts and screws passing through the whole to hold them together. This is not an ideal trough, but it certainly answers its purpose as well as any at present made.

Live-Cages.—These are not used so largely for water objects as for insects, etc. They consist of a brass plate having a glass base-plate, over which a cap slides, having a very thin cover-glass.

The subject to be viewed is placed between these two glasses and held firmly by compression. The best form is that designed by Mr. Rousselet, shown in Fig. 59, with which a condenser may be used conveniently. It is also so arranged that even if a specimen be fixed at the extreme edge of the glass plate, there is room for an objective to work on it. The ordinary live-cages are usually provided with a cover-glass too small in diameter for

this to be done.   A very good plan is often adopted by amateurs
for viewing live objects as follows: A square, flat piece of glass is
obtained, and on this an india-rubber ring is laid, into which the
animalculæ can be placed ; a thin piece of glass is now put over
the top of the india-rubber ring, and this really makes a very
serviceable trough.

Further information on the use of live-cages, compressors, etc.,
will be found on p. 253.

Rousselet's Compressor.—Mr. Rousselet, the designer of the live-

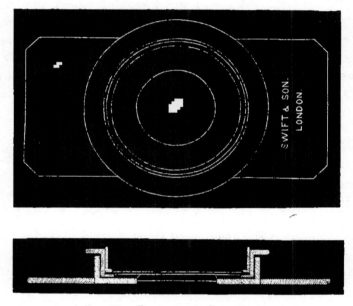

FIG. 59.—ROUSSELET'S LIVE-CAGE.

cage previously mentioned, is also the originator of the most
efficient compressor at present obtainable.   It is shown in
Fig. 60.   The upper arm has cemented to its under side a
portion of a circle of thin cover-glass, which enables high-power
objectives to be employed, and the disc of glass in the base-plate
is not too thick to prevent the employment of condensers of large
aperture.   The compression is quite parallel in action, being
effected by turning a milled head at the top of a drum containing
a spring, which causes the upper plate to rise when the milled
head is released.   The cut-off top of the cover-glass permits of
different media being inserted while the specimen is under

examination, and the arm can be turned aside when desired for cleaning, etc.

Forceps.—Stage forceps are used to hold unmounted specimens in the field of view while they are examined, there being a fitting

on the forceps to go into a hole provided in the limb or the stage of the instrument.

There are in existence many modifications of the apparatus described in the foregoing pages, the adoption or rejection of which must be left

FIG. 60.—ROUSSELET'S COMPRESSORIUM.

to the suggestions which will be naturally derived from practical experience. But the forms of apparatus most commonly worked with, and those whose merits particularly commend them to the writer's judgment, have been described.

Eye-Shade for Monocular Microscope.

It is recommended when working with a monocular microscope that the eye not actually employed should remain open. Many workers experience no difficulty in doing this, but others are quite unable to succeed. For such the eye-shade shown in the accompanying figure (Fig. 61) will be found very advantageous.

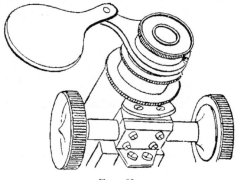

FIG. 61.

It is made of vulcanite, and consists of two pieces jointed in the middle; one side is bored out to the outer diameter of the draw-tube, over which it slides, and the other portion is just a plain piece of sheet vulcanite which obscures light from the disengaged eye. It was introduced by C. Baker.

HINTS REGARDING THE CARE AND USE OF THE MICROSCOPE.

Cleanliness is a most important feature in microscopical work. Never allow dust to accumulate upon the microscope, for it soon finds its way between the fittings, and causes mechanical screws to work with backlash. When not in use, the instrument should always be placed in its case or under a glass shade.

Dusting.—For dusting the microscope, a camel's-hair brush should be used in the first place; by aid of this, dust can be removed from niches and crevices with great ease. For wiping over the stand, also for cleaning eyepiece lenses, the fronts of objectives, and other optical work, the writer has always employed handkerchiefs that are made of a mixture of silk and cotton. These should be washed out two or three times until they are soft and free from dust.

Adjusting a Microscope.—It is not to be recommended that other than microscope makers should take the instruments to pieces; but it is necessary, where a person resides abroad, that he should be able to adjust his own microscope. It is difficult to give definite advice, because the fittings vary considerably in every make of microscope. If the rackwork of the coarse adjustment or sub-stage develop loss of time, it is more often than not due to the bearings clutching on account of the presence of dust, or to their becoming dry. The way to adjust them is as follows : Rack the body up as far as it will go, and mark lines with a pen and ink on the pinion stem and the body of the microscope to correspond with one another. The object of this is to insure the replacement of the body so that the rack engages the correct leaf of the pinion, and it is here presumed that a Jackson model microscope is used, and that it has a stop-pin to the rack, which prevents the body being removed from its fitting. Now remove the cockpiece, which holds the pinion in position, and take away the pinion itself, holding the body meantime, or it will run down on to the stage.

Remove the body from its fitting, wipe both bearings and the rack thoroughly with paraffin oil and a clean rag, then dry them with another cloth. Now drop on, at most, two drops of watchmakers' oil on each side of the bearing fittings attached to the body, and replace the tube in its fittings. It should then be moved up and down until the motion is quite free, and if there are adjusting screws, they should be so set that there shall not be any shake in the fitting of the body, but that it may just, and only just, move in the bearings with its own weight when the instrument is set vertically. Carefully wipe the pinion leaves out, and then, after setting the ink-marks in correspondence again, the pinion may be attached. This usually has adjusting screws to the cockpiece, which push it closer to, or allow it to remain farther from, the rack ; these should be so set as to give a soft movement. It is useless to attempt this procedure with a microscope that never has worked well, but where an instrument has, after use, become unsatisfactory in the mechanical parts, it generally is found to answer. Practically the same treatment is applicable to the mechanical movements of the stage, but very great care is essential, lest either of the plates become bent, an accident that is more easy of occurrence than would be deemed likely.

OBJECTIVES.—It is unwise for unskilled persons to unscrew parts of microscopic objectives ; they are frequently deranged by this means. If at any time it should, from some cause or another, be necessary to unscrew them, an ink-mark or small scratch should be made on each combination, so that when put together again they can be screwed up in the same position as before.

After using an oil immersion objective, the oil must be carefully removed from the front lens by wiping with the handkerchief. Undue pressure must not be used, but it must be thoroughly cleaned. If oil should become dried on the front lens at any time, it will be best to place some fresh immersion oil over it ; it should then be allowed to stand in a place free from dust for about an hour, when the whole may be cleaned off together.

Never use the immersion oil supplied for objectives for lubricating the microscope fittings. It is resinous, and will completely spoil the movements.

Do not touch the microscope with hands soiled by reagents, and care should be taken not to spill any media on the instrument, and particularly the stage.

If dust specks are seen in the field, the eyepiece should be rotated, and if the specks are in it they also will revolve. If they remain stationary they may either be in the objective or in the lamp-glass ; their location can then be ascertained by moving the mirror, when, if the specks move too, they are not in the objective. Dust can be removed from the eyepiece lenses by wiping with a handkerchief dipped in spirits of wine.

Work with the eye close to the eye-lens of the eyepiece to get the best results.

Never use high-power eyepieces when low ones will do.

Be careful not to knock objectives, eyepieces, or condensers, or let them fall.

TRAVELLING.—When travelling with a microscope it is always well to pack the instrument round with tissue-paper, so that it cannot shake in its case. Screws frequently become loosened, and in some instances broken, and movements disordered, by severe shaking while in transit.

For a more exhaustive treatment of the subject the following textbooks are recommended :

Carpenter's 'The Microscope and its Revelations,' edited by the Rev. W. H. Dallinger, LL.D., F.R.S., etc.

'Microscopy,' by Dr. E. J. Spitta.

'Photo-micrography,' by Dr. E. J. Spitta.

'Photo-micrography,' by Mr. Andrew Pringle.

'Practical Photo-Microscopy,' by J. Edwin Barnard, F.R.M.S.

And 'Practical Microscopy,' a handbook for beginners, by Dr. F. Shillington Scales.

As periodicals the best are : *Knowledge*, published at 42, Bloomsbury Square, London, W.C., which devotes special columns to the subject; the *Journal of the Royal Microscopical Society*, published bi-monthly, and the *Journal of the Quekett Club*, published twice a year, both at 20, Hanover Square, London, W.

The Choice of an Outfit.

Summarizing the conclusions arrived at in the foregoing pages, it may be well to briefly recapitulate some points to be borne in mind in selecting an outfit.

The Microscope-Stand.—In the interests of the advancement of microscopy as a science, the best and most suitable means must be commended, and nothing can be gained by encouraging the perpetuation of such instruments as do not embody the accuracy of adjustment or convenience of design which the modern worker with his beautifully perfect optical accessories actually needs in order to derive all the benefit that his lenses are capable of yielding.

It is asserted by users of Continental microscopes, whose name is legion, that the British microscope exceeds the needs of the laboratory worker. The only response from the expert to such a criticism is that there is a want of appreciation and education in matters microscopical in the laboratory. It is impossible to disregard the modern spirit which demands excessive rapidity of work at the cost of excellence and accuracy, and it is not too much to say that the man who examines structures with a good $\frac{1}{12}$-inch oil immersion objective and an Abbe illuminator has limited his knowledge to an extent which would cause him great surprise if he had but the opportunity of seeing the same subject properly illuminated with a sub-stage condenser having a suitable ratio of aperture to that of the objective and the microscope properly manipulated.

Notwithstanding the great strides which have been made in recent years by Continental makers, particularly in the direction of the improvement of the fine adjustment, the instruments of English manufacture alone combine all those refinements which have been found so advantageous by critical workers and have been designed and provided with such excellent judgment. The Continental makers have yet to produce instruments having sub-stages with centring screws, and excepting in two instances, mechanical stages have yet to be built as a part of the instrument, and not as an auxiliary attachment.

Our preference would therefore be for a microscope by one of the well-known English makers, and the following would be the order of preference for the various mechanical fittings :

1. Coarse adjustment by rackwork.
2. Fine adjustment.
3. Compound sub-stage, with screws to centre.
4. Mechanical movements for the stage.
5. Mechanical draw-tube.
6. Fine adjustment to sub-stage.
7. Concentric rotation to the stage.
8. Divided scales, as may be found necessary.
9. Other mechanical fittings, such as centring screws and rackwork to the rotation of the stage, rackwork rotation to sub-stage, etc.

In amplification of the above we would remark that where questions of economy prevail, the sub-stage may be replaced with an under-fitting having centring screws, and the mechanical stage with a sliding-bar.

Many microscopes are made in plain form as a foundation on which, as a superstructure, many of the mechanical fittings can be subsequently mounted. Consideration might with propriety be given by a beginner to such instruments.

On no account purchase a microscope which has not its fittings—that is, the eyepiece, sub-stage, and objective screw of the Royal Microscopical Society's standard gauges.

The Condenser.—Bearing in mind the strictures that have been passed on the Abbe illuminator, it is well on the threshold of work to choose a condenser that will be of permanent utility, and although the Abbe illuminator is sufficiently effective with a $\frac{1}{6}$-inch objective of medium aperture, and with its top lens removed will work with low-power objectives, it is not sufficiently well corrected to develop the possibilities of a $\frac{1}{12}$-inch oil immersion objective. It will always be worth while, therefore, to begin with a well-corrected achromatic condenser.

Objectives.—Having decided whether or no the students' series or the higher class objectives are to be purchased, the powers that should be chosen would depend on the work that was to be undertaken. For general botanical and biological work, 1-inch or $\frac{2}{3}$-inch, and $\frac{1}{6}$-inch are the most useful. In the majority of laboratories the $\frac{2}{3}$-inch is preferred. For bacteriological work a $\frac{1}{12}$-inch oil immersion is added to the above.

The amateur does the greater part of his work with low powers, and he will be well advised to choose at the beginning a 2-inch, 1-inch, and $\frac{1}{6}$-inch.

Eyepieces.—These should not be chosen too high in power, for they will be found of little use. The greater part of the work is done with a No. 2 eyepiece. If a second eyepiece is taken, it should be either a No. 3 or No. 4, the higher powers being only required for testing the quality and performance of objectives.

Sundry Accessories.—All workers will find it an advantage to take a revolving nosepiece to carry two or three objectives, and amateurs will find the inclusion of a bull's-eye condenser for the examination of opaque subjects, troughs for water specimens, together with a live-cage and compressor, useful and advantageous accessories.

CHAPTER V

A SHORT NOTE CONCERNING THE INFLUENCE OF DIFFRACTION ON THE RESOLVING POWER OF MICROSCOPICAL OBJECTIVES, AND ON THE APPARENT COLOUR OF MICROSCOPICAL OBJECTS

By the late Dr. G. JOHNSTONE STONEY, F.R.S.

If we look from a distance at a flame through a thin feather or other uniformly ruled grating we see the flame, and around or on either side of it a number of lateral coloured images, which are wider and usually fainter the farther out that they lie. We thus learn that the light which passes through the grating becomes both a direct beam and a number of lateral, or diffracted, beams, as they are called. The proportions in which the light which passes the grating is distributed between the central beam and the several diffracted beams depends upon the ratio of the widths of the openings to the widths of the bars of the grating, as well as upon such particulars as whether each opening is a mere hole and each bar a mere obstruction, or whether they are occupied by material which acts on the light, especially if it act like a prism. It rarely happens that this distribution does not perceptibly differ for light of different wave-lengths. The direct beam consists of light in very nearly the same state as if it had passed through a simple opening of the size of the grating, except that it is fainter—usually fainter in some colours than in others.

Accordingly, if the eye when looking at the grating, or if the object-lens of a telescope, were to receive only this central beam wherewith to form an image of the grating, the image would be almost identical with that which would be furnished by light

154

coming through an opening covered by tinted glass, and no trace of the ruling would be seen in it. In order to see the ruling, the telescope lens must be able to catch and forward to its focus other rays which have passed through the grating than those of the central beam. The more of the lateral beams which it can transmit and combine at its focus with the light of the central beam (where they will by interference strengthen some parts of the image formed by the central beam, and enfeeble others, thus *introducing* detail), the more nearly will the image it forms resemble the actual grating in detail, and in freedom from false colour. If it succeeds in catching, along with the central beam, even some small portion of the nearest of the diffraction beams, the image will exhibit lines, and the proper number of lines, though it will not present correctly such minuter detail as the widths of the lines and of the spaces between them.

Cases exactly analogous to this occur with the microscope. When an object covered with dots, such as *Pleurosigma angulatum*, has been focussed upon the stage, and is resolved, the diffraction beams may be clearly seen upon the back lens of the objective by removing the eyepiece and looking down the tube. With this diatom there will then be seen the central beam, and portions of the nearest of the lateral beams, six in number. A rather small cone of illumination is best to show them conspicuously if white light be used, and they can be seen with larger cones of illumination and very sharply defined if monochromatic green light, produced by prisms, be employed.

The markings on the *Pleurosigma angulatum* are spaced in each row at intervals which have been measured, and found to be equal to wave-lengths of red light. With so close a ruling the lateral beams are much diffracted or bent aside, and dry objectives can only take in the central beam and a portion of each of the nearest diffracted beams. This enables us to see with such an objective the markings correctly so far as concerns their number and positions, but any further detail is not correctly presented. Immersion objectives can transmit nearly the *whole* of the six nearest lateral beams, which are those that would produce spectra of the first order. We now see some detail: the dots appear hexagonal, and are separated from one another by walls which are thin, and which look like a honeycomb. This is the first and the only step we can take towards learning what the actual detail upon this diatom is, since no objective is competent to supply to the image the second or subsequent diffraction beams ; inasmuch as no immersion fluid can shorten the waves of visible light so much as would enable the object to emit and the lens to receive these further diffraction beams.

The unequal distribution of *colour* between the several beams is strikingly exhibited by the diatom known as *Actinocylus Ralfsii*. The phenomenon may be conveniently examined through a $\frac{1}{2}$-inch apochromatic, over which is mounted an iris diaphragm—an adjunct which is useful for many purposes. Select a frustule which is blue when this upper diaphragm is partly closed. Remove the eyepiece, close the lower, and open the upper diaphragm. Then, on looking down the tube, it will be seen that most of the red is located in the ring of first lateral beams, with, of course, an equal defect of red in the central beam. Hence the blue colour seen when the image is formed by the central beam only. Now place a small central stop (which may be cut out of card) over the back lens of the objective, open the upper, and partially close the lower diaphragm. The image is then formed by the ring of lateral beams only, and will be found to be preponderatingly red.

PART II

CHAPTER VI

INTRODUCTION

IN publishing methods of preparing, staining, hardening, and mounting microscopic objects, I have adopted the system employed in my classes for some years past—that is, each separate stage of procedure is arranged in successive lessons or chapters. A subject such as this cannot be so lucidly described in writing as by demonstration, but it has been my aim to make it as clear as possible, so that if the instructions are carefully followed and practised, successful permanent work can be performed; but it is only by most scrupulous care and constant practice that any degree of success in this work can be attained.

In books on this and cognate subjects it too often happens that tools, instruments, and routine are prescribed that tend to make work needlessly laborious and expensive, and are in consequence causes of discouragement to the readers. The directions given in the succeeding pages will, it is believed, commend themselves for their directness and simplicity. They are, moreover, thoroughly practical, and are the processes that I have found the most effective after more than thirty years' experience as a mounter of microscopic objects.

137

HARDENING AND PRESERVING ANIMAL TISSUES AND LISTS OF MATERIALS.

LIST OF TISSUES AND ORGANS, AND THE MOST SUITABLE HARDENING, STAINING, AND MOUNTING REAGENTS.

TISSUES.	HARDENING REAGENT.	STAINING FLUID.	MOUNTING MEDIUM.	ANIMAL.
Bld, ...	Dry on slide.	Eosin.	C. Balsam.	Frog.
Do., amphibia.	Do.	Eosin and methyl green.	Do.	Do.
E...	2 per ... bichromate potash.		F ...'s.	
Endothelium.	Nil.	Silver ni...	Farrant's or C. Balsam.	Rabbit.
White fibrous tissue.	Alcohol.	Nil.	F ...'s or glycerine jelly.	Tail of rat.
Yellow elastic tissue.	Chromic acid and spirit.	Nil.	do.	Lig. nuchæ of ox.
Adipose tissue.	Methylated spirit.	Hæmatoxylin.	do.	... animal.
Tendon.	Do.	Do.	do.	Sheep.
Adenoid tissue.	Müller's fld.	Do., and eosin.	C. Balsam.	Do.
Cartilage	Chromic acid and spirit.	Do. do.	Do.	Do.
Bone.	Chromic and nitric acid.	Picrocarmine.	Farrant's.	Do.
Do., developing.	Do. do.	Hæmatoxylin and ...	C. Balsam.	Kitten.
Marrow.	Methylated spirit.	Hæmatoxylin.	Do.	Guinea-pig or cat.
M..., striated.	2 per cent. ... potash.	Do., and ...	Do.	Cat.
Do., non-striated.	...ic acid and spirit.	Hæmatoxylin.	Balsam or Farrant's.	Kid of rabbit.
Nerve-fibres.	Osmic acid.	Osmic acid.	Farrant's.	Sciatic of frog.
Do., trunk.	Chromic acid and spirit.	Hæmatoxylin and eosin.	C. Balsam.	Do. of cat.
B...	Do. do.	Do. do.	Do.	Cat.
Lymphatic glands.	Müller's fld.	Do. do.	Do.	Do.
Tonsil.	Methylated spirit.	Do. do.	Do.	Do.
Thymus gland.	Müller's fld.	Do. do.	Do.	Human fœtus or calf.

LIST OF TISSUES AND ORGANS, AND THE MOST SUITABLE HARDENING, STAINING, AND MOUNTING REAGENTS—*continued.*

TISSUES.	HARDENING REAGENT.	STAINING FLUID.	MOUNTING MEDIUM.	ANIMAL.
Skin.	Methylated spirit.	Hæmatoxylin and in.	C. Balsam.	Human palm of hand.
Nail.	Do. do.	Do. do.	Do.	Human toes.
Scalp.	Do. do.	Do. la.	Do.	Do.
Heart	Chromic acid and spirit.	Do. la.	Do.	Cat.
Trachea.	Do. do.	Do. lo.	Do.	Do.
Lung.	Do. do.	Do. lo.	Do.	Do.
Tooth.	Chromic and nitric acid.	Picrocarmine.	F Do.	Do.
Do., developing.	Do.	Do.	Do.	Kitten about two months old.
Tongue.	Chromic acid and spirit.	Hæmatoxylin and eosin.	C. Balsam.	Cat.
Œsophagus.	Do. do.	Do. do.	Do.	Do.
Stomach, cardiac end.	Absolute alcohol.	Safranine and	Do.	Do.
Do., pyloric end.	Do. do.	Hæmatoxylin and in.	Do.	Do.
Small intestine.	Chromic acid and spirit.	Do. d.	Do.	Do.
Large intestine.	Do. d.	Do. lo.	Do.	Do.
Liver.	2 per cent. bichromate of	Do. do.	Do.	Do.
Pancreas.	Absolute alcohol.	Do. do.	Do.	Do.
Salivary glands.	Do. do.	Do. do.	Do.	Do.
Spleen.	2 per cent. bichromate potash.	Do. do.	Do.	Do.
Supra-renal glands.	Methylated spirit.	Do. d.	Do.	Guinea-pig or cat.
Thyroid glands.	Do. d.	Do. do.	Do.	Do.
Kidney.	2 per cent. bichromate potash.	Do. d	Do.	Do.
Ureter.	Chromic acid and spirit.	Do. do.	Do.	Do.
Testicle.	Methylated spirit.	Do. do.	Do.	Do.

LIST OF TISSUES AND ORGANS, AND THE MOST SUITABLE HARDENING, STAINING, AND MOUNTING REAGENTS—*continued.*

TISSUES.	HARDENING REAGENT.	STAINING FLUID.	MOUNTING MEDIUM.	ANIMAL.
Vas deferens.	Methylated spirit.	Hæmatoxylin and eosin.	C. Balsam.	Cat.
Epididymis.	Do.	Do. do.	Do.	Do.
Bs.	Do.	Do. do.	Do.	Do.
Ovary.	Do.	Do. do.	Do.	Do.
Fallopian tube.	Do.	Do. d.	Do.	Do.
bd.	Do.	Do. do.	Do.	Do.
Mammary gland.	Do.	Picrocarmine.	Farrant's or C. Balsam.	Do.
Spinal cord.	2 gr cent. bichromate ammonium.	blue-b la.k.	C. Balsam.	Do.
Me oblongata.	Do. do.	Do. d.	Do.	Do.
Pons Ba.	Do. do.	Do. do.	Do.	Do.
id.	Do. do.	Do. d.	Do.	Do.
Cerebrum.	Do. la.	Do. do.	Do.	Do.
Ej.	Methylated spirit.	Hæmatoxylin and in.	Do.	Cat or human fœtus.
ba.	Müller' fluid.	Do. d.	Do.	Cat or sheep.
Gd.	Do. do.	Do. do.	Do.	Sheep.
Crystalline l ns.	2 gr nt. b' ImBate aph.	Picrocar ns.	F ns's.	ld.
Retina.	Bb' fluid.	Hæmatoxylin and in.	C. Balsam.	Ox.
Sclerotic	Do. do.	Do. do.	Do.	Bd.
Optic ne.	Do. do.	Do. do.	Bd.	Do.
Olfact ay m us membrane.	Do. do.	D.. do.	Do.	Rat.
Internal ear. Ch-lea.	Müller's fluid, and de-fgd.	Carmine in bulk.	Do.	Guinea-pig.

LIST OF BOTANICAL SPECIMENS, AND THE MOST SUITABLE PRESERVING, STAINING, AND MOUNTING MEDIA.

SPECIMEN.	PRESERVING REAGENT.	STAINING FLUID.	MOUNTING MEDIUM.
Stems, young.	Me th...yed spirit.	Hæmatoxylin.	C. Balsam.
Do., older.	Do. do.	Carmine and acid green.	Do.
L ...	Do. do.	Hæmatoxylin.	Do.
Ovaries.	Do. do.	Do.	Do.
Anthers.	Do. do.	Borax carmine.	Do.
Epidermis for stomata.	Macerate in water.	Mhyl aniline.	Glycerine jelly.
Fibro-vascular tissues.	Do. do.	Acid aniline green.	Do. do.
Yeast.	Camphor-water.	Unstained.	Camphor-water.
Green algæ.	Acetate of copper solution.	Do.	...te of copper solution.
Red algæ.	Dilute ...ted spirit.	Do.	Glycerine jelly.
Protococcus.	Acetate of copper solution.	Do.	Acetate of copper solution.
Volvox.	Do. do.	Do.	Do. do.
D ...ls.	Do. do.	Do.	Do. do.
Raphides.	Macerate in water.	Do.	C. Balsam.
...et.	Methylated spirit.	Do.	Glycerine jelly.
Fertile branch of ...	Do. do.	Do.	Camphor-water.
Antheridia and archegonia of mosses.	Do. do.	Do.	C. Balsam.

CHAPTER VII

HARDENING AND PRESERVING ANIMAL TISSUES AND ORGANS FOR MICROSCOPICAL EXAMINATIONS

FRESH untreated tissues are unsuited for microscopical purposes, but it is sometimes advisable to observe the appearance of some specimens, such as muscle-fibres, tendon, connective tissues, and nerve-fibres, while fresh. When this is desired, the tissue must be examined in certain fluids called 'normal fluids,' that will alter its character as little as possible. Those generally used are: (1) Blood serum; (2) the aqueous humour from a fresh eye; and (3) normal or ¾ per cent. salt solution. The two former are difficult to obtain, but the latter can be made at any time, and it will answer for most purposes. Place a small piece of the tissue on a slide, add a drop or two of salt solution, take two needles fixed in holders and carefully separate the fibres from each other; this process is called teasing. When sufficiently teased, apply a cover-glass and examine. You may now wish to irrigate with some staining reagent; if so, place a few drops of the stain at one edge of the cover-glass, and apply a piece of blotting-paper to the other side; this will absorb the salt solution, and the staining fluid will follow and take its place around the tissue; the slide may then be placed under the microscope, and the action of the reagent observed.

These specimens cannot, as a rule, be kept. For permanent preparations the tissues or organs must be hardened. This is accomplished by subjecting them to the action of certain hardening or fixing solutions. The following are most commonly used:

Absolute Alcohol.—Suitable for stomach, pancreas, and salivary glands. These organs must be perfectly fresh, and

142

they should be cut into small pieces, so that the alcohol may penetrate as quickly as possible.

Change the alcohol every day for the first three days. The hardening is usually complete in a week.

Chromic Acid and Spirit.—Chromic acid ⅙ per cent., watery solution 2 parts, and methylated spirit 1 part. This reagent hardens in about ten days. Then transfer to methylated spirit, which should be changed every day until no colour comes away from the tissues. It is suitable for cartilage, nerve-trunks, heart, lips, bloodvessels, trachea, lung, tongue, bladder, ureter, intestines, and œsophagus.

Potassium Bichromate.—Make a 2 per cent. watery solution. This will harden in about three weeks. Then transfer to methylated spirit, and change the spirit every day until no colour comes away from the tissues. It is suitable for muscle, spleen, liver, and kidney.

Ammonium Bichromate.—Make a 2 per cent. watery solution. It hardens in from three to four weeks. Then transfer to methylated spirit, and change every day until no colour comes away from the tissues. It is suitable for spinal cords, medulla, pons Varolii, cerebellum, and cerebrum.

Müller's Fluid.—Bichromate of potash 30 grains, sulphate of soda 15 grains, distilled water 3½ ounces. It hardens in from three to five weeks. Then transfer to methylated spirit, and change every day until no colour comes away from the tissues. Suitable for lymphatic glands, eyeballs, retina, and thymus gland.

Methylated Spirit.—May be used universally, if preferred, but it has a tendency to shrink some tissues too much. It hardens in about ten days. Change the spirit every twenty-four hours for the first three days. Suitable for skin, scalp, testicle, penis, prostate gland, vas deferens, epididymis, ovary, uterus, Fallopian tubes, placenta, mammary gland, supra-renal glands, tonsils, and all injected organs.

Decalcifying Solution.—For bones. Make a ⅙ per cent. watery solution of chromic acid, and for every ounce add 5 drops of nitric acid. This fluid will soften the femur of a dog in about three weeks; larger bones will take longer. Change the fluid several times, and test its action by running a needle through the thickest part of the bone. If it goes through easily, the bone is soft enough ; if not, continue the softening process a little longer. When soft enough, transfer to water, and soak for an hour or two ; then pour off the water and add a 10 per cent. solution of bicarbonate of soda, and soak for twelve hours to remove all trace of acid. Wash again in water, and place in methylated spirit until required. Bones and teeth should always be softened in a large quantity of the decalcifying solution.

Olfactory Region.—Divide with a saw the head of a freshly killed rabbit or guinea-pig longitudinally, and parallel with the nasal septum. Cut out the septum so as to expose the olfactory region, which is recognized by its brown colour. Dissect out a portion including some of the turbinated bones. Harden this in Müller's fluid for three or four days. Then transfer to chromic and nitric acid decalcifying solution, and soak until the bones are quite soft. Wash well in water to remove all trace of acid, and complete the hardening in methylated spirit.

Cochlea.—Dissect out the internal ear of a freshly killed young guinea-pig, open bulla with bone-forceps, when a conical elevation, the cochlea, will be seen. Remove as much of the surrounding bone as possible, and place the cochlea in Müller's fluid for two weeks to harden the delicate nervous tissues. Then transfer to chromic and nitric acid decalcifying solution, and soak until the bone is soft. Place in weak spirit for a day or two, and then transfer to strong methylated spirit.

Corrosive Sublimate.—Tissues may be fixed very quickly in corrosive sublimate. Make a saturated solution in 5 per cent. glacial acetic acid. The specimens should be removed from the solution as soon as they are fixed, directly they become opaque throughout. Then wash in repeated changes of 70 per cent. alcohol, to which a little tincture of iodine has been added. This process will fix tissues in a few minutes.

Picric Acid.—Make a saturated solution in water. This solution will fix small pieces of tissue in a few minutes ; larger specimens will require from three to six hours' immersion. Then wash out the picric acid with repeated changes of spirit. Water must not be used, as it is hurtful to the tissues that have been prepared by this method. For the same reasons, during all subsequent stages of treatment, water should be avoided, and the staining should be carried out in alcoholic solutions.

Formaldehyde.—This may be used universally if desired. It is sold commercially as ' formal ' in a 40 per cent. solution. This must be reduced by the addition of water to a 2 or 4 per cent. solution. It is specially useful for hardening nervous tissues and for eyes ; the latter are completely hardened in twenty-four hours.

When in great haste, tissues may instantly be fixed in boiling water.   Boil some water in a test-tube, then drop in small pieces of the tissue, and boil again for a few seconds.   The specimen may then be placed at once in gum and syrup, and when penetrated, freeze, and make the sections.   This method should only be used when a section is urgently wanted.

General Directions for Hardening Tissues.

1. Always use fresh tissues.
2. Cut the organs into small pieces with a sharp knife.
3. Never wash a specimen in water ; when it is necessary to remove any matter, allow some normal salt solution to flow over the surface of the tissue, or wash in some of the hardening reagent you are going to use.
4. All specimens should be hardened in a large quantity of the reagent ; too many pieces should not be put into the bottle, and they should be kept in a cool place.
5. In all cases the hardening process must be completed in spirit.
6. Label the bottles, stating the contents, the hardening fluid used, and when changed.   Strict attention to these details is necessary for successful histological preparations, for if the hardening is neglected good sections cannot be made.

CHAPTER VIII

EMBEDDING TISSUES AND SECTION-CUTTING

To Cut Sections with a Razor by Hand.—Take the tissue between the thumb and forefinger of the left hand. Hold the finger horizontally, so that its upper surface may form a rest for the razor to slide on. Take the razor, hold it firmly in the hand, keep the handle in a line with the blade, and draw it through the tissue from heel to tip towards yourself. While cutting, keep the razor well wetted with dilute methylated spirit, and as the sections are cut place them in a saucer of dilute methylated spirit.

Embedding in Paraffin Wax and Lard.—Melt together by the aid of gentle heat four parts of solid paraffin and one part of lard. A quantity of this may be made and kept ready for use at any time. Melt the paraffin mass over a water-bath. Take the specimen and dry it between the folds of a cloth to remove the spirit, so that the paraffin may adhere to its surfaces, place it in a pill-box in the desired position, and pour in enough melted paraffin to cover it, then set aside to cool. When quite cold, break away the pill-box and cut sections from the embedded mass with a sharp razor. When a number of specimens are embedded, and it is desired to keep them for some time, they should be preserved in a jar of methylated spirit.

To Infiltrate a Tissue with Paraffin.—Dehydrate the specimen in absolute alcohol for several hours, then transfer to chloroform or xylol, in which it must remain until perfectly saturated. When clear, place in a bath of melted paraffin of 45° C. (melting-point), and keep it at this temperature for several hours, so that the paraffin may penetrate to the middle of the tissue. Then remove it from the paraffin and put it into a small pill-box, pour

146

in enough paraffin to fill the box, and as the paraffin cools, add a little more to make up the shrinkage and set aside to cool. When cold, place in water for a few minutes; this will soften the paper, and facilitate the removal of the pill-box. You will now have a cylinder of paraffin with the specimen firmly fixed in its centre, and, if desired, the paraffin may be pared away from the sides until a square block is obtained. The sections may now be made by hand with a razor, or the block can be fixed to a microtome with a little melted paraffin.

The sections must be placed in turpentine to remove the paraffin, then in absolute alcohol to remove the turpentine, and finally, in distilled water to remove the alcohol; they may then be stained. Sometimes it is desirable to stain the tissue in bulk before it is embedded. In this case the sections need only go into turpentine or benzole to wash away the paraffin; they may then be mounted in Canada balsam.

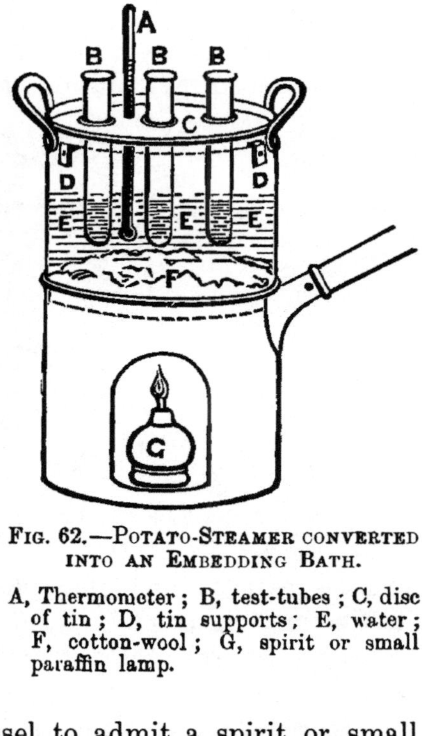

The above process requires an embedding bath. **This is** usually an expensive affair, but one that will answer all ordinary purposes can easily be made.

Fig. 62.—Potato-Steamer converted into an Embedding Bath.

A, Thermometer; B, test-tubes; C, disc of tin; D, tin supports; E, water; F, cotton-wool; G, spirit or small paraffin lamp.

Get a small potato-steamer, and cut a hole in the lower vessel to admit a spirit or small paraffin lamp. Get a tinsmith to cut out a circular plate of tin to fit into the upper vessel, in which some holes must be cut to take the test-tubes, and to the sides of the vessel four small pieces of tin, bent at right angles, must be soldered to support the tin plate. A piece of tin must also be soldered over the perforated bottom of the vessel, so that it will hold water. When the alterations are complete, place a layer of cotton-wool or a piece of felt on the bottom of the steamer, to protect the test-tubes from breakage; half fill with water, add a thermometer, light the lamp, and on the desired temperature being attained, put some paraffin in the test-tubes, place them in the steamer, and when the paraffin has melted add the specimens.

After use dry the apparatus so that rust may not set in. If this is attended to it will last for years.

When a proper embedding bath cannot be obtained, tissues may be infiltrated with paraffin in the following way : Dehydrate the specimen in absolute alcohol ; then place in a quantity of chloroform or benzole, ten or twelve times the bulk of the tissue, until saturated ; add small pieces of paraffin until no more will dissolve, and set aside for several 'hours. Apply gentle heat to drive off the solvent and melt the paraffin, after which the tissue

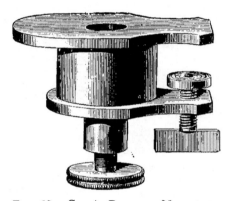

Fig. 63.—Cole's Pattern Microtome.

can be removed and embedded in a pill-box of paraffin of the desired melting-point.

Cole's Microtome and Embedding in Carrot.—When a number of sections are wanted, or when a complete section of an organ is desired, a microtome should be used. A very good and simple instrument can be obtained from Messrs. Watson and Sons, 313, High Holborn. Screw the microtome firmly to the table, and with the brass tube supplied with the microtome punch out a cylinder of carrot to fit into the well of the microtome. Cut this in half longitudinally, and scoop out enough space in one half of the carrot to take the specimen ; then place the other half of carrot in position, and make sure that the specimen is held firmly between them, but it must not be crushed. Now put the cylinder of carrot and specimen into the well of the microtome and commence cutting the sections. A good razor will do, but it is better to use the knife which Messrs. Watson supply with the microtome. While cutting, keep the knife and plate of the microtome well wetted with dilute methylated spirit, and as the sections are cut place them in a saucer of dilute spirit. A number of sections may be cut and preserved in methylated spirit until required.

When a specimen has a very irregular outline, it cannot be successfully embedded in carrot. Paraffin should then be used. Place the tissue in the well of the microtome in the desired position, pour in enough melted paraffin to cover it, and when cold cut the sections.

Freezing Microtome.—Cathcart's is the most simple and cheapest freezing microtome, and it can be obtained from any optician.

1. Cut a slice of the specimen about ⅛ inch thick, in the direction you wish to make the section.

2. Place in water for an hour to remove the alcohol.

3. Transfer to a mixture of gum-water 5 parts, saturated watery solution of loaf-sugar 3 parts, and allow it to soak in this for about twelve hours; or, if a few drops of carbolic acid are added to the mixture, tissues may remain in it for months without harm.

4. Clamp the microtome to a table, fix the ether spray in its place, and fill the bottle with ether. Methylated ether, specific gravity 720, will do.

5. Put a little gum and syrup on the zinc plate of the microtome, and place the tissue in it. Commence working the bellows, and as soon as all the gum has frozen add some more and freeze again, and so on until the tissue is completely covered and frozen into a solid mass. This proportion of gum and syrup works well in a temperature of 60° F.; but when higher, less syrup is required—when lower, more. The syrup is used to prevent freezing too hard, so some judgment must be exercised in the matter.

6. The best instrument for making the sections is the blade of a carpenter's plane. Hold it firmly in the right hand, and work the microtome screw under the machine with the left. Plane off the sections as quickly as possible. They should all collect

on the plane iron.   If they roll up or fly off, the tissue is frozen too hard, or there is not enough syrup in the gum.   If the former is the case, allow the mass to thaw a little ; if the latter, add some more syrup to the gum mixture, and soak the tissue again.

When the sections are cut, place them in a saucer of water, which must be changed several times until all trace of gum is

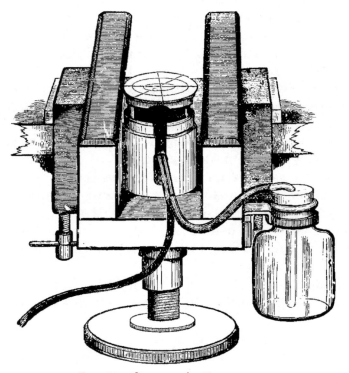

Fig. 64.—Cathcart's Microtome.

removed.   Water that has been boiled and allowed to cool will remove the gum sooner than cold water.   When quite free from gum, the sections may be bottled up in methylated spirit until required for staining.

Embedding in Celloidin.—Dissolve Schering's celloidin in equal parts of absolute alcohol and ether until the solution is as thick as glycerine.   Divide the solution into 2 parts, to 1 of which add an equal part of absolute alcohol and ether.   De-hydrate the specimen in absolute alcohol for several hours, then transfer to the thinner solution of celloidin, and soak until perfectly saturated ; place in the thick celloidin for about an hour, or until required.   Take a cork and paint over one end a layer of celloidin, and let it dry ; this will prevent air-bubbles rising from the cork and lodging in the mass.   Take the specimen from the celloidin and lay it on the cork, and let it stand for a minute or two, then add some more celloidin until the tissue is completely covered, and set aside, and when the mass has attained such a consistency that on touching it with the finger no impression will remain, place it in 50 per cent. alcohol for an hour or two to complete the hardening, or it may remain there until required.   The embedded mass can now be placed between two pieces of carrot, and put into an ordinary microtome, and the sections made with a knife or razor, which must be well wetted with methylated spirit ; or the embedded specimen can be removed from the cork, and, after soaking in water, it can be transferred to gum and syrup, and the sections made with a Cathcart freezing microtome.   If it is desired to remove the celloidin from the sections, soak them in equal parts of absolute alcohol and ether.   When all the celloidin is removed, transfer to distilled water, then into the stain.   After staining, wash in distilled water, dehydrate, clear in clove oil, and mount in Canada balsam.

When it is not desirable to remove the celloidin from the sections, they should be stained in borax carmine or hæmatoxylin. The former stains celloidin, but the colour is removed by washing in acidulated alcohol.   Hæmatoxylin only stains it slightly.   All the aniline dyes stain it deeply ; they should not be used.

Tissues are usually stained in bulk before they are infiltrated with celloidin.   When so, the sections must be dehydrated in methylated spirit, cleared in oil of bergamot or origanum, and mounted in Canada balsam.

When desirable, sections infiltrated with celloidin may be mounted in Farrant's medium or glycerine jelly.   Wash away all trace of alcohol with water, and mount in either of the above media in the ordinary way.

Celloidin is an excellent medium for infiltrating many specimens of both animal and vegetable subjects.   The following are a few of these :

Flower-buds of lily, yucca, evening primrose, poppy, dandelion, and anthers ; worms, leech, flukes, gills and organs of mussels, heads of frogs, newts, sponges, etc.

For flower-buds proceed as follows : Harden the bud in methylated spirit in the ordinary way. Then take a piece of fine silk or cotton and tie it round the centre of the bud to hold the parts together ; now with a sharp knife cut off each end of the bud so that the celloidin may easily penetrate to the interior. Now place the specimen in equal parts of absolute alcohol and ether for at least twelve hours. Then transfer to the thin solution of celloidin, and soak until completely infiltrated. Remove and place in thick celloidin for about twelve hours. Take out of celloidin on the point of a needle, and hold exposed to the air for a few minutes, to dry the celloidin around the exterior of the bud. When dry, push gently off the needle into some methylated spirit, and soak for at least twelve hours to complete the hardening of celloidin. The specimen may then be embedded in carrot, and the sections may be cut in any ordinary well microtome. Worms must be cut up into pieces of about $\frac{1}{4}$ or $\frac{1}{2}$ inch long ; these are then dehydrated in equal parts of ether and alcohol, infiltrated with and embedded in celloidin, and then treated in exactly the same way as directed for flower-buds.

When a number of celloidin masses are prepared for future use, they must be preserved in a vessel of methylated spirit.

Embedding in Gelatine.—This method is very useful for hairs, cotton, silk, wool, and all such fibres. Take, for example, some human hairs about $\frac{1}{2}$ inch long, and make a bundle of them ; tie them together either with a long hair or with some fine cotton. Place the bundle in warm water and soak for a few minutes. Now make up a strong solution of some clear transparent gelatine. Cox's is very good—say 1 ounce of gelatine to 6 of water. Transfer the bundle of hair to this, place in a warm water-bath, and soak until the gelatine has penetrated all through the bundle. Remove from gelatine on the point of a needle, and allow the mass to cool ; then place in methylated spirit for about twelve hours. The embedded mass may then be placed in a cylinder of carrot and transverse sections cut in the ordinary well microtome. The sections when cut are to be placed in strong spirit to dehydrate; they are then cleared in clove oil and mounted in Canada balsam.

Heads of frogs, newts, and many other specimens may be infiltrated and embedded in gelatine, but they must all be stained in bulk before they are infiltrated, because the sections must not come in contact with water in any form; moreover, if the sections were stained the gelatine would be coloured as well as the tissues.

The Rocking Microtome.—This machine is made by the Cambridge Scientific Instrument Company.   It is only used for specimens infiltrated with paraffin, and it is automatic—that is to say, it can be set to cut sections of definite thickness, and

Fig. 65.—Rocking Microtome.

every time the handle is pulled a section is cut, and the specimen is moved forward ready for another.

Infiltrate the tissue with paraffin in the ordinary way in a pill-box, and when the paraffin has set, remove the box and trim the paraffin into a rectangular block.   Take care to keep the edges quite parallel, so that they may adhere together, as the sections are cut and form a riband.   The Cambridge Instrument Company make an apparatus for embedding, called embedding L's.   If these are used, perfectly rectangular blocks are formed ready for fixing to the brass cap at the end of the arm of the microtome, which is filled with paraffin; this should be warmed over a spirit-lamp, and the block containing the specimen is to be pressed against the melted paraffin until it adheres firmly.

CHAPTER IX

STAINING ANIMAL SECTIONS AND MOUNTING IN CANADA BALSAM

ALL sections of organs and tissues should be stained with some colouring reagent, so that their structure may be made more apparent. Certain parts of the tissue have a special affinity for the dyes or stain; they therefore become more deeply tinted, and stand out clearly from the surrounding tissues.

The following staining reagents are the most useful:

Grenacher's Alcoholic Borax Carmine.—Carmine, 3 grammes; borax, 4 grammes; distilled water, 100 c.c. Dissolve the borax in the water, add the carmine, and apply gentle heat until all is dissolved; then add 100 c.c. of 70 per cent. alcohol, filter, and keep in a stoppered bottle.

Staining Process.—1. Place the section in distilled water to wash away the alcohol, then place a little of the carmine in a watch-glass, and immerse the section for from three to five minutes.

2. Wash the section in methylated spirit.

3. Take of methylated spirit 5 parts, and of hydrochloric acid 1 part, and mix them well together. A quantity of this acid solution may be made up and kept ready for use at any time.

Immerse the section in the above, and leave it to soak for about five to ten minutes, or, if overstained, until the desired tint is obtained. Sections of skin and scalp may be left until all colour is removed from the fibrous tissues; the glands, hair follicles, and Malpighian layer will then stand out clearly.

4. Wash the section well in methylated spirit to remove all traces of the acid, then transfer to some perfectly clean and

154

strong methylated spirit for from ten to fifteen minutes to dehydrate.

5. Place some oil of cloves in a watch-glass, take the section from the spirit on a lifter, and carefully float it on to the surface of the oil, in which it must remain for about five minutes. This process is called clearing; the object of it is to remove the alcohol and to prepare the section for the balsam.

6. Transfer the section to some filtered turpentine to wash away the oil of cloves, and mount it in Canada balsam. Sections may be mounted in Canada balsam direct from the oil of cloves, but it is better to wash in turpentine first, because if much oil is mixed with the balsam it will not dry; the oil also has a tendency to cause the balsam to turn a dark yellow colour.

Ehrlich's Hæmatoxylin.—Hæmatoxylin, 30 grains; absolute alcohol, $3\frac{1}{2}$ ounces; distilled water, $3\frac{1}{2}$ ounces; glycerine, $3\frac{1}{2}$ ounces; and ammonia alum, 30 grains. Dissolve the hæmatoxylin in the alcohol and the alum in the water; mix the two solutions together, and add the glycerine and 3 drachms of glacial acetic acid. The mixture must now be left exposed to light for at least a month, then filter and keep in a stoppered bottle.

Staining Process.—1. If the specimen has been hardened in any of the chromic solutions, place the section in a 1 per cent. watery solution of bicarbonate of soda for about five minutes, then wash well in distilled water. If it is a spirit preparation the soda will not be required, but all sections must be washed in distilled water before they go into hæmatoxylin stain.

2. To a watch-glassful of distilled water add from 10 to 20 drops of the hæmatoxylin solution, and immerse the section for from ten to thirty minutes.

3. Wash in distilled water, then in ordinary tap water; the latter will fix the dye and cause the colour to become blue.

When a section has been overstained with hæmatoxylin, the excess of colour may be removed by soaking it for a few minutes in a $\frac{1}{2}$ per cent. solution of glacial acetic acid in distilled water, then wash again in tap water.

4. Dehydrate in methylated spirit.

5. Clear in clove oil, and mount in Canada balsam.

Double Staining with Hæmatoxylin and Eosin.—Stain the section in hæmatoxylin, as directed above, then place it in an alcoholic solution of eosin—about 1 grain of eosin to 1 ounce of methylated spirit is strong enough—and let it soak for about five minutes; wash well in methylated spirit, clear in clove oil, and mount in Canada balsam.

Ehrlich's hæmatoxylin is a good all-round stain, but as it is acid it must not be used for tissues of a mucous nature, such as the umbilical cord, and many tumours containing mucous or gelatinous tissues. Delafield's hæmatoxylin should then be used.

To 400 c.c. of a saturated aqueous solution of ammonia alum add 4 grammes of hæmatoxylin dissolved in 25 c.c. of absolute alcohol; leave the solution exposed to the light and air in an unstoppered bottle for three or four days; filter, and add to the filtrate 100 c.c. of glycerine and 100 c.c. of wood spirit (methylic alcohol); allow the solution to stand in the light until it becomes of a dark colour, refilter, and keep in a stoppered bottle.

Use as directed for Ehrlich's stain.

Aniline Blue-Black.—Dissolve 30 grains of nigrosine in $3\frac{1}{2}$ ounces of distilled water, then add 1 ounce of rectified alcohol and filter. This stain is only used for sections of brain and spinal cord. Immerse the sections for from thirty to sixty minutes, wash in water, dehydrate in methylated spirit, clear in clove oil, and mount in Canada balsam.

Aniline Blue.—Make a 1 per cent. solution of soluble aniline blue in distilled water and filter. Stain the section for five to ten minutes, wash in water, and place in methylated spirit, in which it must soak until the excess of colour is removed. Clear in clove oil and mount in Canada balsam.

This stain is useful for cardiac glands of the stomach, brain, and spinal cord.

Golgi's Nitrate of Silver Methods.—These are chiefly employed for investigating the relations of cells and fibres in the central nervous system. Two methods are mostly used, as follows:

(a) Very small pieces of the tissue, which have been hardened for some weeks in bichromate solution or Müller's fluid, are placed for half an hour in the dark in 0·75 per cent. nitrate of silver solution, and are then transferred for twenty-four hours or more to a fresh quantity of the same solution (to which a drop or two of formic acid may be added). They may then be hardened with 50 per cent. alcohol, and sections, which need not be thin, are cut either from celloidin with a microtome or with the free hand. The sections are mounted in Canada balsam, which is allowed to dry on the slide. They must not be covered with a cover-glass, but the balsam must remain exposed to the air.

(b) Instead of being slowly hardened in bichromate, the tissue is placed at once in very small pieces in a mixture of bichromate and osmic (3 parts of Müller's fluid to 1 of osmic acid). In this it remains from two to five days, after which the pieces are treated with silver nitrate, as in the other case. This method is not only more rapid than the other, but is more sure in its results.

Mounting in Canada Balsam.—Take 3 ounces of dried Canada balsam and dissolve in 3 fluid ounces of pure benzol, filter, and keep in an outside stoppered bottle. Clear the section in clove oil, and place in turpentine. Clean a cover-glass and a slide, place a few drops of balsam on the centre of the latter, take the section from the turpentine on a lifter, allow the excess of turpentine to drain away, and with a needle-point pull the section off the lifter into the balsam on the slide. Now take up the cover-glass with a pair of forceps, and bring its edge in contact with the balsam on the slide ; ease it down carefully, so that no air-bubbles are enclosed, and with the points of the forceps press on the surface of the cover until the section lies quite flat, and the excess of balsam is squeezed out. The slide must now be put aside for a day or two to allow the balsam to harden ; the exuded medium may then be washed away with some benzol and a soft camel's-hair brush, after which dry the slide carefully with a cloth and apply a ring of cement. The above method answers well for mounting sections quickly, but when time will admit the following is a much better way : Clear the section and place it in turpentine ; clean a cover-glass, and moisten the surface of a slide with your breath ; apply the cover-glass to the slide, and make sure that it adheres. Place a few drops of balsam on the cover, into which put the section. Now put the slide away in a box, or in some place out of reach of dust, for twelve hours, so that the benzol may evaporate from the balsam. Clean a slide, warm it gently over the flame of a spirit-lamp; apply a drop of balsam to the surface of the hardened balsam on the cover-glass; take the cover up in a pair of forceps, and bring the drop of fresh balsam in contact with the centre of the warmed slide. Ease the cover down carefully, so that no air-bubbles may be enclosed, press on the surface of the cover-glass until the section lies quite flat; set the slide aside to cool. The exuded balsam may then be washed away with methylated spirit and a soft rag, and a ring of cement applied.

Staining in Bulk. — Place small pieces of the tissue in Grenacher's alcoholic carmine for from one to three days, then transfer to a $\frac{1}{2}$ per cent. solution of hydrochloric acid in methylated spirit for from one to twelve hours, according to the size of the tissue. Wash well in spirit, and soak for a day in 90 per cent. spirit.

The specimen may then be infiltrated and embedded in paraffin, celloidin, or gelatine, but be careful to follow the instructions previously given with each method.

Flemming's Method for Staining Karyokinetic Nuclei. — Fix the tissue in the following Flemming's solution:

Osmic acid, 1 per cent. solution ...	80 c.c.
Chromic acid, 10 per cent. solution	15 c.c.
Glacial acetic acid	10 c.c.
Distilled water	95 c.c.

The fixing process is usually complete in twelve hours; then wash the tissue thoroughly in water and harden in alcohol of gradually increasing strength. Now place small shreds or thin sections in a saturated alcoholic solution of saffranin mixed with an equal quantity of aniline water for two days. The tissue is then to be washed in distilled water. It is then soaked in absolute alcohol until the colour is removed from everything except the nuclei. It is then again rinsed in water and placed in a saturated watery solution of gentian violet for two hours, washed again in distilled water, decolorized in alcohol until only the nuclei are left stained; then transfer to bergamot oil and mount in xylol balsam.

Weigert-Pal Method for the central nervous system, by which all medullated fibres are stained darkly, while the grey substance and any sclerosed tracts of white matter are left uncoloured. Pieces of brain or spinal cord which have been hardened in Müller's fluid are to be placed direct in gum-water and syrup and soaked for a few hours; then make sections with a freezing microtome, and place them in water, and from this transfer to Marchi's fluid, as follows:

Müller's fluid	...	...	... ...	2 parts,
Osmic acid, 1 per cent.	...			1 part,

and soak for a few hours. They are then washed in water and transferred to the following stain: Dissolve 1 gramme of hæmatoxylin in a little alcohol, and add to it 100 c.c. of a 2 per cent. solution of glacial acetic acid, in which leave the section for twelve hours; it will then be quite black. Wash again in water, and place in a ¼ per cent. solution of potassic permanganate for five minutes; rinse with water and transfer to Pal's solution (sulphate of soda, 1 gramme; oxalic acid, 1 gramme; distilled water, 200 c.c.), and bleach for a few minutes. When sufficiently bleached they are passed through water into alcohol, cleared in bergamot oil, and mounted in Canada balsam.

Ehrlich's Triple Stain for Blood-Corpuscles.

Saturated watery solution, orange ' G '	...			135 parts.	
..	,,	,,	methyl green	...	110 ,,
..	..	,,	acid fuchsin		100 ,,

To the above add

Glycerine	...	...	...	...	100 parts.
Absolute alcohol	...	..	...	...	200 ,,
Distilled water	...	...	...	...	300

This solution should stand for several weeks to allow for sedimentation, and it improves with age. When used the supernatant liquid should be drawn off with a pipette to avoid the sediment.

The cover-glasses are to be well cleaned with alcohol, and the surface of one is touched with a drop of fresh blood, and another cover-glass pressed on its surface until the blood is evenly distributed. The covers are then separated and allowed to dry. When dry they must be still further hardened over a spirit-lamp, or on a hot stage made of sheet copper, and kept at 212° F. for from fifteen minutes to two hours; after which place in stain for from one to four minutes, wash in water, dry, and mount in Canada balsam, benzol, or xylol.

The eosinophile granules in the corpuscles will be a reddish hue, the neutrophile granules purple, and the nuclei bluish-green or blue.

Toison's Solution for Staining White Blood-Corpuscles.

Methyl violet ...	...	...	...	½ grain.
Neutral glycerine	...	...	...	1 ounce.
Distilled water ...	...	...	...	2½ ounces.

Mix thoroughly and add—

Chloride of sodium	...	...	...	15 grains.
Sulphate of sodium	...	...	...	2 drachms.
Distilled water ...	...	...	...	5½ ounces.

Filter and keep in a stoppered bottle. Spread blood on cover-glass, dry, and immerse in stain for eleven minutes. Wash in water, dry and mount in Canada balsam.

Fixing and Staining Sections on the Slide.

Mayer's Albumen Method.—White of egg, 50 c.c.; glycerine, 50 c.c.; salicylate of soda, 1 gramme: shake well together, and filter into a stoppered bottle. A thin layer of the cement is spread on a slide with a brush, and the section laid on it. Now warm gently on a water-bath. As the paraffin melts it is carried away from the section by the albumen. The section may now be washed with turpentine, benzole, and alcohol, and be treated with aqueous or other stains, without fear of it moving.

Shellac Method.—Make a solution of shellac in absolute alcohol —it should be about the thickness of oil—filter, and keep in a stoppered bottle. Warm some slides, and spread over them a layer of the cement with a brush, and put away to dry. When dry apply a very thin layer of creosote; this will form a sticky surface, on which the section must be carefully laid. Now heat the slide on a water-bath for about fifteen minutes at the melting-point of the paraffin; this will allow the section to come down on the shellac film, and at the same time evaporate the creosote. Allow the slide to cool, and wash away the paraffin with turpentine or benzol. If the section has been stained in bulk, a drop or two of Canada balsam is added, and a cover-glass applied.

To Stain a Section on the Slide.—Fix section on slide as directed above. Wash away the paraffin with rectified mineral naphtha, follow this quickly with a few drops of methylated spirit, and then with some distilled water. Now apply the stain, and place the slide under a bell-glass to prevent evaporation; or the slide may be plunged into a vessel containing the staining solution. When sufficiently stained, wash with distilled water, dehydrate with methylated spirit, drain away the spirit, and apply a drop of clove oil to clear the specimen. When clear, drain away as much of the oil as possible, add a drop of Canada balsam, and apply the cover-glass.

CHAPTER X

STAINING BLOOD AND EPITHELIUM, TEASING-OUT TISSUES, AND MOUNTING IN AQUEOUS MEDIA STAINING WITH PICROCARMINE, GOLD CHLORIDE, SILVER NITRATE, AND OSMIC ACID

Double Staining Nucleated Blood-Corpuscles.

STAIN A.—Dissolve 5 grains of eosin in ½ ounce of distilled water and add ½ ounce of rectified alcohol.

Stain B.—Dissolve 5 grains of methyl green in an ounce of distilled water.

Place a drop of frog's blood on a slide, and with the edge of another slide spread it evenly over the centre of the slip ; now put it away out of reach of dust to dry. When quite dry, flood the slide with Stain A for three minutes. Then wash with water, and flood the slide with Stain B for five minutes. Wash again with water, and allow the slide to dry. Apply a drop or two of Canada balsam and a cover-glass.

Blood of Mammals, Non-Nucleated Corpuscles.

Spread a drop of blood on a slide and let it dry for twelve hours, then stain in a strong alcoholic solution of eosin for about five minutes, drain away the eosin, rinse the slide in methylated spirit, let it dry, apply a drop of Canada balsam and the cover-glass.

Both of the above processes should be carried out during dry weather, as any moisture in the air retards the drying of the corpuscles, and then they are liable to change their form.

Epithelium.—Kill a frog, cut off its head, and remove the lower jaw. Open the abdomen and take out the stomach, and slit it open. Place the head, lower jaw, and stomach in a 2 per cent. solution of bichromate of potash for forty-eight hours. Then wash gently in water until no colour comes away from the specimens. Now place all three portions in picrocarmine for twenty-four hours. Remove the tissues from the carmine, and allow the stain to drain away from them. Take the lower jaw and scrape the tongue for squamous epithelium, and place the deposit obtained in a few drops of glycerine on a slide. Take the stomach, remove some columnar epithelium from its internal surface, and place it in some glycerine on another slide. Then take the head for ciliated epithelium, which will be found at the hinder part of the roof of the mouth ; put some scrapings from this in glycerine on a slide as before. Clean a slide and place a drop or two of Farrant's medium on its centre ; take up a little of the epithelium on the point of a needle, and put it into the medium. Now apply a cover-glass, and with the needle-point press it down until the epithelial cells are separated and spread evenly between the cover and the slide. Set the slide aside for a day or two, so that the medium may set. Then wash away the excess of medium with some water and a camel's-hair brush, dry the slide with a soft rag, put it in a turn-table, and run on a ring of cement.

Portions of the tongue, trachea, and intestine of a rabbit or cat may be treated in the same way.

Endothelium.—Take a piece of the omentum of any small animal, and rinse gently in distilled water to remove soluble matter. Place it in a $\frac{1}{4}$ per cent. solution of silver nitrate for ten minutes, or until it becomes a milky white. Wash well in ordinary water, and expose in a saucer of water to diffused sunlight, until it assumes a brownish colour. Cut out a small piece and mount it in Farrant's medium or glycerine jelly. In this specimen only the interstitial cement substance will be seen. To compare with it, cut out a similar piece, wash it in distilled water, and stain it with hæmatoxylin for ten minutes ; wash away all excess of stain with distilled water, and mount in Farrant's medium or glycerine jelly. In this specimen the nuclei will be seen stained blue. Specimens of mesentery showing endothelium may also be mounted in Canada balsam. When this is desired, stain the tissue as directed above, dehydrate in methylated spirit, clear in clove oil, and mount in Canada balsam.

Teasing-out Tissues.—Take a very small piece of the tissue, place it on a slide in a few drops of distilled water, and with a couple of needles mounted in holders carefully separate the fibres from each other. When the parts are sufficiently isolated, drain away the water, add a few drops of the mounting fluid, and apply the cover-glass. When teasing it is very important that a proper background should be used so that the object may be easily seen. For a coloured specimen, a piece of white paper should be used, and a transparent white tissue will be seen better on a dark ground, such as a piece of black paper or American cloth ; the slide should be examined from time to time under the low power of the microscope to ascertain when the tissue is teased out enough.

White Fibrous Tissue.—Harden some tendons from a rat's tail in methylated spirit for a week. Then soak a small piece in water to remove all trace of spirit, place it on a slide in a few drops of water, and tease it up until the fibres are separated from each other. Drain away the water, add some Farrant's medium or glycerine jelly, and apply a cover-glass.

Yellow Elastic Tissue.—Place small pieces of the ligamentum nuchæ of an ox in chromic acid and spirit for ten days. Then proceed as above.

Striped or Voluntary Muscle.—Harden small pieces of muscle of a pig in a 2 per cent. solution of bichromate of potash for three weeks, then transfer to methylated spirit, in which it may remain until required. Soak a piece in water to remove the spirit, place a very small fragment on a slide in a few drops of water, and with a couple of needles tease or tear the tissue up so as to separate the fibres. Drain away the excess of water, apply a drop or two of Farrant's medium or glycerine jelly and a cover-glass.

Non-Striped Muscle.—Harden a piece of the intestine of a rabbit in chromic acid and spirit for ten days. Wash in water, strip off a thin layer of the muscular coats, and stain it in hæmatoxylin. Wash in distilled water, and then soak in ordinary tap-water until the colour becomes blue. Clean a slide, pass a small fragment of the muscle on it in a few drops of water, and with needles separate the fibres.   Drain off the excess of water, apply a few drops of Farrant's medium or glycerine jelly and a cover-glass.

Nerve-Fibres.—Dissect out the sciatic nerve of a frog, and stretch it on a small piece of wood as follows : Take a match, make a slit in each end of it, into which put the ends of the nerve ; now place it in a 1 per cent. solution of osmic acid for an hour or two.   Wash in water, tease up a small fragment on a slide, and apply a few drops of Farrant's medium or glycerine jelly and a cover-glass.

When staining with gold chloride, solutions from ½ per cent. to 5 per cent. in distilled water are employed.   It is used for staining nerves and nerve-endings ; it also brings out the cells of the cornea, fibrous connective tissues, and cartilage.

The tissue must be taken from the animal immediately after death, and be placed in the solution of gold for from half an hour to an hour ; it is then removed to distilled water for twelve hours, and afterwards exposed to the action of diffuse sunlight in a saturated solution of tartaric acid or formic acid until it assumes a purple colour.

The future treatment will depend on the nature of the specimen.

If muscle has been stained for nerve-endings, place a small piece on a slide, tease it up, and examine with a low power until you find a nerve-fibre terminating in an end-plate on a muscle-fibre, separate it from the surrounding fibres as much as possible, add some Farrant's medium or glycerine jelly, and apply a cover-glass.

If cornea or cartilage, make vertical and horizontal sections with a freezing microtome, and mount in Farrant's medium or glycerine jelly.   Sections of gold-stained tissues may also be mounted in Canada balsam ; when this is desired, dehydrate in strong spirit, clear in clove oil and mount in Canada balsam.

There are many ways of staining with gold, but the above is the most simple, and it gives very good results.   For the other methods the student may refer to the larger works on practical histology.

Staining with Picrocarmine.—Rub up 1 gramme of carmine with 10 c.c. of water, and 3 c.c. of strong liquid ammonia ; add this to 200 c.c. of a saturated solution of picric acid in distilled water. Leave the mixture exposed to the air until it evaporates to one-third of its bulk ; filter, and keep in a stoppered bottle. Place some of the picrocarmine in a watch-glass, and immerse the section for from half an hour to an hour. Remove from the stain with a lifter, and place the section on a slide ; drain away as much of the excess of stain as possible, and, if necessary, soak up what remains with a piece of filter-paper. Then add a few drops of Farrant's medium, and apply the cover-glass.

Picrocarmine stained tissues should never be washed ; if they are, all the yellow colour will be removed, and the specimen will come out stained with carmine only. They improve by keeping, and the staining process goes on for several days after they are mounted ; that is to say, some parts give up the stain, and others absorb it. Picrocarmine may be purchased in crystals, with which a 2 per cent. solution in distilled water should be made.

If it is desirable to mount a picrocarmine stained section in Canada balsam proceed to stain as above ; then make a saturated solution of picric acid in methylated spirit, filter, and dehydrate the section in it ; then give it a final rinse in methylated spirit, clear in clove oil and mount in Canada balsam.

Farrant's Medium.—Take of glycerine and a saturated aqueous solution of arsenious acid equal parts, and mix them well together; then add as much powdered gum arabic as the mixture will take up, and let it stand for six weeks. Filter, and keep in an outside stoppered bottle.

The above is difficult to make; it is better to obtain it ready for use.

Glycerine Jelly.—Dissolve 1 ounce of French gelatine in 6 ounces of distilled water ; then melt in a water-bath, and add 4 ounces of glycerine and a few drops of creosote or carbolic acid. Filter through paper while warm, and keep in a stoppered bottle. The above may be used instead of Farrant's medium. The jelly must, of course, be warmed before use. All tissues or sections must be well soaked in water before they are mounted in Farrant's medium or glycerine jelly, so that all trace of alcohol is removed.

Tissues containing much air should be soaked in water that has been boiled for about ten minutes and allowed to cool.

CHAPTER XI

STAINING AND MOUNTING MICRO-ORGANISMS

THE investigation of bacteria may be carried out under various conditions:

(1) In fluids, such as milk, water, blood, pus, etc. (2) On solid media, bread, meat, potatoes, meat jelly, etc., or in the tissues and organs of animals. In the former case a drop of fluid is placed on the centre of a cover-glass, and another cover-glass is placed on it; the two glasses are then to be rubbed together to spread the organisms evenly over their surfaces; they are then separated and allowed to dry. When bacteria are growing on solid material, scrape off a small portion, put on a cover-glass, and treat as above; separate the covers, and allow to dry. When the cover is quite dry, take it up with a pair of forceps, organisms uppermost, and pass two or three times through the flame of a spirit-lamp; this will fix the albumen and fasten the bacteria to the glass.

To Stain Bacteria on Cover-Glasses.—They should be floated with the organisms downwards on a saturated watery solution of any of the following aniline dyes: Methyl blue, methyl violet, gentian violet, fuchsin, vesuvin, or Bismarck brown. From ten to fifteen minutes is enough for the first four stains; vesuvin and Bismarck brown require about an hour. When the staining is complete wash the cover in distilled water. If the colour is too deep wash it in a ½ per cent. solution of acetic acid, and then again in water; put away to dry. When quite dry add a drop of Canada balsam, and mount on a slide in the usual way.

When bacteria are present in the organs of animals the tissues should be hardened in methylated spirit for about a week, and very thin sections with a freezing microtome cut from them. The sections may be stained in any of the above dyes; then wash in water, dehydrate in absolute alcohol, clear in oil of cedar or bergamot, and mount in balsam.

Staining Bacillus Tuberculosis.

Ehrlich's Method for Double Staining.—To 100 parts of a saturated watery solution of aniline oil add 11 parts of a saturated alcoholic solution of fuchsin, and filter. Place the covers or sections in the stain in a watch-glass, and warm slowly over a spirit-lamp until vapour rises. Wash in water, and then immerse for about a minute in dilute nitric acid, 1 part of acid to 2 parts water. Wash again in water, and stain again in a solution of methyl blue—100 parts of distilled water to 20 parts of a saturated solution of methylated blue—in alcohol for about twenty minutes. Wash in water, and in the case of sections dehydrate in absolute alcohol, clear in oil of cedar or bergamot, and mount in balsam. The cover-glass preparations must be dried; then add a drop or two of balsam, and mount as above.

Ziehl Neelsen's Method.—Fuchsin, 1 part; 5 per cent. watery solution of carbolic acid, 100 parts; absolute alcohol, 10 parts. Remove the section from the alcohol, and immerse in the above stain for fifteen minutes.

Decolorize in a 5 per cent. watery solution of sulphuric acid, wash well in water to remove acid, and counter-stain in the following for five minutes: Saturated alcoholic solution of methyl blue, 1 c.c.; distilled water, 5 c.c. Wash in water, dehydrate in absolute alcohol, clear in cedar oil, and mount in Canada balsam.

Gibbe's Double Stain. — Rose aniline hydrochloride, 2 grammes; methyl blue, 1 gramme; rub well together in a mortar. Then dissolve aniline oil, 3 c.c., in 15 c.c. of rectified spirit, and add the crystals to the mixture; shake well, and when all are dissolved add 15 c.c. of distilled water. Place the cover-glass preparation or sections in the stain in a watch-glass, and warm gently over a spirit-lamp; then let them soak for four or five minutes. Wash in methylated spirit until no colour will come away, clear in oil of cedar, and mount in Canada balsam.

Cover-glass preparations will not require clearing; they are allowed to dry, then add a drop of balsam and mount on a slide.

Anthrax Bacillus.

Löffler's Alkaline Blue Method.—To 100 parts of a solution of caustic potash (1 in 10,000) in distilled water, add 30 parts of a saturated alcoholic solution of methylene blue. Immerse the sections for an hour. Wash in distilled water, and then in a ½ per cent. solution of acetic acid in distilled water. Wash away the acid with water, dehydrate in alcohol, clear in oil of cedar or bergamot, and mount in Canada balsam.

Anthrax bacilli may also be stained by Gram's method.

Gram's Method.—*Solution A.*—Saturated alcoholic solution of gentian violet, 11 parts; saturated watery solution of aniline, 100 parts. Mix well together, and filter.

Solution B.—Iodine, 1 part; iodide of potassium, 3 parts; distilled water, 300 parts.

Solution C.—Saturated aqueous solution of vesuvin.

Take the section from alcohol, and place in Solution A for one to three minutes. Wash in alcohol, and transfer to Solution B for three minutes. Wash in alcohol, and place in Solution C for five minutes. Wash in distilled water, dehydrate, clear in oil of cedar or bergamot, and mount in Canada balsam.

Leprosy Bacillus.—Stain in the following for three minutes ·

Fuchsin ...	...	...	...	1 gramme
Rectified spirit ...				20 c.c.
Distilled water ...		...	...	80 c.c.

then place for thirty seconds in a solution consisting of 90 per cent. alcohol 10 parts, and nitric acid 1 part. Wash in water, dehydrate in absolute alcohol, clear in cedar oil, and mount in Canada balsam.

Diphtheria Bacillus.—Employ Löffler's alkaline blue and proceed as for anthrax.

Glanders Bacillus.

Kuhne's Method.—Methylene blue, 1 gramme; absolute alcohol, 10 c.c. When all the blue has dissolved, add 100 c.c. of a 5 per cent. watery solution of carbolic acid.

The sections are transferred from alcohol to the above stain for half an hour. Wash in water and place in a weak solution of acetic acid in distilled water until they are of a pale blue colour; watch carefully, or too much colour may be removed; they are then rinsed in lithia water (1 in 70) 1 c.c., water 34 c.c., and transferred to water. The sections are now to be taken up one at a time on the point of a needle and dipped into absolute alcohol, in which some methylene blue has been dissolved. Dehydrate in methylene aniline oil, made as follows: Rub up about 10 grammes of methylene blue with 10 c.c. of aniline, and let the mixture settle. When dehydrated, rinse in aniline, and place for a few minutes in terebene to clear, then mount in Canada balsam.

Schutz's Method.—Stain the sections or cover-glass films in methylene blue, 1 gramme; rectified spirit, 20 c.c.; distilled water, 80 c.c., for several hours. Wash in a ½ per cent. solution of acetic acid, dehydrate in absolute alcohol, and clear in cedar oil, and mount in Canada balsam.

Syphilis Bacillus.—Stain by Lusgarten's method as follows:

Aniline oil	3 c.c.
Distilled water	100 c.c.
Saturated alcohol solution of gentian violet	11 c.c.
Alcohol	10 c.c.

Sections or cover-glasses are placed in the above for from twelve to twenty-four hours. They are then transferred to absolute alcohol for a few minutes; then place for ten seconds in a 1 per cent. solution of permanganate of potassium, and wash in 5 per cent. solution of sulphuric acid to decolorize the ground tissue. Wash in water, dehydrate in absolute alcohol, clear in cedar oil, and mount in Canada balsam.

Bacillus of Enteric Fever—Gaffky's Method.—Sections or cover-glasses are placed for twenty-four hours in a strong solution freshly made by adding a saturated alcoholic solution of methylene blue to distilled water.   They are then washed in distilled water, dehydrated in absolute alcohol, cleared in terebene, and mounted in Canada balsam.

Spirillum.—On cover-glasses these are easily stained by any aniline dye solutions.   When in sections use the following stain for twenty-four hours :

Bismarck brown ...	1 gramme.
Rectified alcohol	20 c.c.
Distilled water	80 c.c.

Then wash in water, dehydrate in absolute alcohol, and mount in Canada balsam.

Actinomycosis.

Stain the sections in the following for ten minutes, warmed to about 45° C.: Magenta, 2 parts ; aniline oil, 3 parts ; rectified spirit, 20 parts ; distilled water, 20 parts.   Wash in water. Place in a concentrated alcoholic solution of picric acid for five to ten minutes.   Wash in water, dehydrate in alcohol, clear in clove oil, and mount in Canada balsam.

Weigert's Method.—Glacial acetic acid, 5 c.c. ; absolute alcohol, 20 c.c. ; distilled water, 40 c.c. ; add orseille until a dark red fluid is obtained.   Stain the sections in the above for an hour ; rinse quickly in alcohol.   Clear in cedar oil and mount in Canada balsam.

Hæmatozoa of Laveran.—Touch a drop of blood with a perfectly clean cover-glass, apply another cover-glass, press them gently together, then slide them apart, and dry.   Now stain in an alcoholic solution of methylene blue, wash in water, dry, and mount in Canada balsam.

Filaria.—Specimens are obtained by pricking the finger of the patient and applying a drop or two of blood to a glass slide ; spread evenly with the aid of a thin glass rod, and allow it to dry. Now apply a few drops of Ehrlich's hæmatoxylin, and stain for about five minutes.   Wash in distilled water, then stain again in a solution of eosin in alcohol, rinse in water, let the slide stand up on end to drain and dry, and then apply a drop or two of Canada balsam on the cover-glass.

Vermes.—Sections of specimens such as Acaris, Tænia, etc., may be made in the following way : Harden the worm in alcohol for a week or ten days. Then cut up in pieces of about ¼ inch long, and soak in equal parts of ether and alcohol for twelve hours ; they are then transferred to a thin solution of celloidin in equal parts of ether and alcohol, and must remain in this until perfectly infiltrated. Now remove from thin celloidin and place in a thicker solution, and soak again for twelve hours. Remove from celloidin on the point of a needle, and hold exposed to the air for a minute or two so that the celloidin may dry all round the exterior of the specimen ; then push it off the needle into methylated spirit, in which it should remain for twelve hours to complete the hardening of celloidin in the interior. Cut transverse sections, and stain in borax carmine for five minutes. Wash in methylated spirit, and then place in acidulated spirit—1 part hydrochloric acid in 5 of methylated spirit—for about three minutes if overstained, until the excess of stain is removed. Wash again well in methylated spirit to remove all trace of acid. Then transfer, for about one to two minutes, to absolute alcohol, clear in oil of origanum, and mount in Canada balsam.

Great care must be taken not to leave the section in absolute alcohol for more than two minutes, or the celloidin will be dissolved and the section will fall to pieces.

Heads and segments of tape-worms, flukes, etc., may all be mounted whole. Harden in methylated spirit for a few days, then stain in borax carmine for from one hour to twenty-four hours according to the size of the specimen. Wash in methylated spirit, and soak in acidulated spirit until the excess of stain is removed. Then place in water for a few minutes to soften the tissue a little. Place the specimen on a glass slide, put another slide on it, and press down carefully until quite flat. Now bind the two slides together with twine and place in a jar of methylated spirit, and soak for at least twenty-four hours. Then remove the twine, separate the slides carefully, and place the specimen in absolute alcohol for ten minutes to dehydrate.   Clear in clove oil for one hour, and mount in Canada balsam.

Anchylostoma.—Harden in methylated spirit for ten days; then stain in borax carmine, wash in methylated spirit, and place in acidulated alcohol to remove excess of colour.   Transfer to water, and soak until all trace of spirit is removed; then mount in glycerine jelly.

Trichina Spiralis.—Harden muscle with trichina encysted in methylated spirit.   Then embed in celloidin, make longitudinal sections, stain in borax carmine, pass through acidulated spirit; then wash in water, and mount in glycerine jelly.

These worms may also be isolated from the muscle.   Tear out a piece of muscle on a glass slide with the aid of a dissecting microscope or pocket lens, separate the capsule containing the trichina from the muscle with a needle, and place it in dilute hydrochloric acid until the capsule is dissolved and the worm is set free; then pick it up with a fine sable brush, wash in water, and mount in glycerine jelly.

CHAPTER XII

INJECTION OF BLOODVESSELS

Carmine and Gelatine Injecting Mass.—

Pure carmine ...	... 60 grains.
Liq. ammonia fort.	... 2 drachms.
Glacial acetic acid	... 86 minims.
Gelatine solution (1 ounce in 6 ounces	
of water)	... 2 ounces.
Water	... 2 ounces.

Dissolve the carmine in the ammonia and water in a test-tube, and mix it with one-half of the warm gelatine. Add the acid to the remaining half of gelatine, and drop it little by little into the carmine mixture, stirring well all the time with a stick or glass rod. Filter through flannel, and add a few drops of carbolic acid to make the mass keep. The principle to be remembered in making this mass is this : the carmine, if alkaline, would diffuse through the vessels and stain the tissues around them ; if acid, the carmine would be deposited in fine granules, which would block up the capillaries ; hence the necessity for a *neutral* fluid. The best guides are the colour and smell of the fluid. It should be a bright red, and all trace of smell of ammonia must be removed. The gelatine solution is made by putting 1 ounce of gelatine into 6 ounces of water ; it must then be left until the gelatine becomes quite soft ; then dissolve over a water-bath.

Prussian or Berlin Blue and Gelatine Mass.—Take 1½ ounces of gelatine, place it in a vessel and cover it with water ; allow it to stand until all the water is absorbed and the gelatine is quite soft. Then dissolve in a hot-water bath. Dissolve 1 drachm of Prussian or Berlin blue and 1 drachm of oxalic acid in 6 ounces of water, and gradually mix it with the gelatine solution, stirring well all the time ; then filter through flannel.

Watery Solution of Berlin Blue.—Dissolve 2½ drachms of the blue in 18 ounces of distilled water, and filter. This fluid is useful for injecting lymphatics.

Injecting Apparatus Required.—An injecting syringe fitted with a stop-cock, and several cannulæ of various sizes.

Directions for Injecting.—The animal to be injected should be killed by chloroform, so that the vessels may be dilated, and injected while warm ; if possible it should be placed in a bath of water at a temperature of 40° C. Expose the artery of the parts to be injected, clear a small portion of it from the surrounding tissues, and place a ligature of thin twine or silk round it. With sharp scissors make an oblique slit in the wall of the vessel, insert the cannula, and tie the ligature firmly over the artery behind the point of the cannula, into which put the stop-cock. Fill the syringe with injection-fluid, which must not be too warm, and take care not to draw up any air-bubbles ; now insert the nozzle of the syringe into the stop-cock and force in a little fluid ; remove the syringe, so that the air may escape, insert the syringe again, and repeat the process until no air-bubbles come out of the stop-cock. You may then proceed slowly with the injection. Half an hour is not too long to take over the injection of an animal of the size of a cat. The completeness of an injection may be judged by looking at the vascular parts, such as the tongue, eyelids, and lips. When the injection is complete shut the stop-cock, remove the syringe and cannula, and tie the ligature round the artery. Now place the animal in cold water for an hour to set the injection-fluid. When quite cold, dissect out the organs, cut them up into small pieces, and place them in methylated spirit to harden, and change the spirit every twenty-four hours for the first three days. The hardening will be complete in ten days.

Injection of Lymphatics (*Puncture Method*).—A small subcutaneous syringe is filled with a watery solution of Berlin or Prussian blue, and the nozzle is thrust into the pad of a cat's foot. The injection is to be forced into the tissues. Then rub the limb from below upwards. This will cause the injection-fluid to flow along the lymphatics and find its way into the glands of the groin.

To Inject **Lymph-Sinuses of Glands.**—Force the nozzle of a subcutaneous syringe into the hilum of a lymphatic gland of an ox, and inject a watery solution of Prussian or Berlin blue until the blue appears on the surface of the gland. Then place it in methylated spirit to harden.

When *blue* injection-fluid is used, add a few drops of acetic acid to the spirit while hardening the tissues.

CHAPTER XIII

CUTTING, STAINING, AND MOUNTING VEGETABLE SECTIONS

STEMS, leaves, roots, etc., should be hardened in methylated spirit for a week or ten days, and the spirit changed every twenty-four hours for the first three days. The stems must not be too old. One, two, and three years' growth will show all that is required.

Wheat, barley, maize, peas, etc., are usually obtained dry. They must be placed in water for a few hours or until they resume their natural shape. Then lay a piece of blotting-paper on a plate, moisten it with water, and spread a layer of the grains on its surface ; now place another piece of wet blotting-paper over all, and put in a warm place for from twelve to twenty-four hours, so that the embryo may begin to germinate. Then remove from the plate, and place the grains in a bottle of methylated spirit, which must be changed every day until all trace of water is removed. The specimens may then be section-ized, or they may remain in spirit until required.

Ovaries.—Gather some before the flower opens, and others after it has been open for a day. You will then have the ovules in both stages. Place them in methylated spirit and change every twenty-four hours for the first three days.

Anthers.—Treat in exactly the same way as ovaries, but anthers must be infiltrated with celloidin before the sections can be cut. Remove the ends, place in equal parts of alcohol and ether, and soak for twelve hours ; then place in celloidin, and, after soaking for from twelve to twenty-four hours, proceed as directed in Chapter VIII. on Section-Cutting.

Some specimens after being in spirit are too hard to cut easily. They may be softened by soaking in warm water. Leaves are often particularly troublesome in this respect ; they bend and become fixed by the action of the spirit, and will not then stand the slight pressure required to hold them firmly between the carrot without cracking. When this happens, soak the leaf in warm water until it is quite pliable ; it can then be embedded in carrot without any risk of being broken. Stems and petioles of many palms are naturally too hard, and they may contain a large amount of silica. They must be soaked in water for a while ; then transfer to liq. potassæ for from one to twelve hours. Wash again well in water to remove all trace of potash, then reharden in methylated spirit. The shells of many stone fruits may be softened and cut by this method.

Section-Cutting, by hand and with a microtome, should be done in the same manner as described in Chapter VIII.

Bleaching.—Vegetable sections generally require bleaching before they can be properly stained. Chlorinated soda is used for this purpose. Take of dry chloride of lime, 2 ounces ; of washing soda, 4 ounces ; and distilled water, 2 pints. Mix the lime in one pint of the water and dissolve the soda in the other. Mix the two solutions together, shake well, and let the mixture stand for twenty-four hours. Pour off the clear fluid, filter, and keep in a stoppered bottle in a dark place, or cover the bottle with paper. Soak the sections in distilled water. Pour off the water and add a quantity of bleaching fluid. Allow this to act for from one to twelve hours. Wash well in water, which must be changed several times to remove all traces of soda. The sections may now be stained, or they may be preserved in spirit until required.

Staining Borax Carmine (suitable for ovaries, fruits, etc.).— Pure carmine, 1 drachm ; liq. ammoniæ fort., 2 drachms. Dissolve the carmine in the ammonia, and 12 ounces of a saturated solution of borax in distilled water. Filter and keep in a stoppered bottle.

1. Put some stain in a watch-glass, and immerse the section for three to five minutes.

2. Wash well in methylated spirit.

3. Take of hydrochloric acid, 1 part ; and of methylated spirit, 5 parts. Mix well together, and soak the section until the colour changes to a bright scarlet, which takes about five

minutes. The acidulated spirit may be kept ready for use at any time.

4. Wash well in methylated spirit. Then place in some strong methylated spirit, and soak for at least ten minutes to dehydrate.

5. Place the section on the surface of a small saucer of clove oil, and let it soak until clear.

6. Remove from the clove oil and place in turpentine, and then mount in Canada balsam.

Full instructions for mounting in Canada balsam are given at end of chapter.

Hæmatoxylin.—Hæmatoxylin, 30 grains; absolute alcohol, $3\frac{1}{2}$ ounces; distilled water, $3\frac{1}{2}$ ounces; glycerine, 3 ounces; ammonia alum, 30 grains; glacial acetic acid, 3 drachms. Dissolve the hæmatoxylin in the alcohol and the alum in the water; then add the glycerine and acetic acid. Mix the two solutions together and let the mixture stand for at least a month before use.

1. Add about thirty drops of the above to an ounce of distilled water, and stain the section for fifteen to thirty minutes.

2. Wash well in distilled water, and then in ordinary tap water. This will fix the colour and make it deeper.

3. Dehydrate in strong methylated spirit for at least ten minutes.

4. Clear in clove oil and mount in Canada balsam.

Double Staining—Dalton Smith's Method.—Stems, roots, and leaves :

> *Green Stain.*—Acid aniline green ... 2 grains.
> Distilled water 3 ounces.
> Glycerine 1 ounce.

Mix the water and glycerine together, and dissolve the green in the mixture.

> *Carmine Stain A.*—Borax 10 grains.
> Distilled water 1 ounce.
> Glycerine $\frac{1}{2}$ ounce.
> Alcohol rect. $\frac{1}{2}$ ounce.

Dissolve the borax in the water, and add the glycerine and alcohol.

Carmine Stain B.—Carmine   10 grains.
     Liq. ammoniæ   20 minims.
     Distilled water   30 minims.

Dissolve the carmine in the water and ammonia. Mix A and B together, and filter.

1. Place the section in green stain for five to ten minutes.
2. Wash in water.
3. Place in carmine from ten to fifteen minutes.
4. Wash well in methylated spirit.
5. Dehydrate and clear in clove oil. Wash in turpentine and mount in Canada balsam.

Double Staining—M. J. Cole's Method.—

Pure carmine ... ... 1 drachm.
Liq. ammonia ... ... 2 drachms.

Dissolve the carmine in the ammonia and add 12 ounces of a saturated solution of borax in distilled water. Filter through paper and keep in a stoppered bottle.

Bleach the sections, and after being well washed with repeated changes of water they are placed in the above stain for five to ten minutes. Then wash well in methylated spirit, and soak in acidulated alcohol—1 part hydrochloric acid to 35 of methylated spirit—until the excess of stain is removed; about two minutes is usually sufficient. Wash again in methylated spirit to remove all trace of acid. Dissolve 5 grains of acid aniline green in 6 ounces of methylated spirit, and filter if necessary. Soak the section in this green stain for at least half an hour; then just rinse in methylated spirit, clear in clove oil, and mount in Canada balsam. The advantage of this method is that the section can remain in the green stain for any time. The writer keeps a stock of sections in it ready for mounting. Should a specimen be overstained green, the excess of colour can easily be removed by soaking in methylated spirit for a few minutes.

Staining with Eosin.—Make a 2 per cent. solution of eosin in alcohol, filter if necessary, and keep in a stoppered bottle. This stain is used for showing the structure of sieve-tubes and plates; it stains protoplasm deeply. Make transverse and longitudinal sections of the stem of a vegetable marrow, and immerse them in the above for ten minutes. Then wash out any excess of colour with methylated spirit, clear in clove oil, and mount in Canada balsam.

Staining Hairs on Leaves.—Make a 2 per cent. aqueous solution of soluble aniline blue, and filter. Now take, for example, a young leaf of *Deutzia scabia*, cut it into small pieces of about ¼ inch square, and bleach in chlorinated soda. Then wash well in water, and immerse in the above stain for twelve hours, wash well in water, and transfer to methylated spirit, in which they must be soaked until nearly all the colour is removed. Then soak in clove oil for several hours, and when quite clear mount in Canada balsam.

Leaves of eucalyptus and other plants showing essential oil glands may be treated in the same way, but if the specimens have been preserved in spirit they must be soaked in water before the bleaching process.

Male and Female Conceptacles of Fucus and other Algæ.—Place the specimens in methylated spirit, which must be changed every twenty-four hours for the first three days, then let them soak for ten days or until required for cutting into sections.

Embed a conceptacle in carrot, place in microtome, and make transverse sections, which must be as thin as possible. While cutting, keep the knife well wetted with methylated spirit, and, as the sections are cut, put them into spirit; no water must come near them. When ready, stain the sections in a strong solution of acid aniline green in spirit for several hours. Then just rinse in absolute alcohol, clear in clove oil, and mount in Canada balsam.

Transverse and longitudinal sections of the thallus of an alga may be treated in the same way, but they may be mounted without staining, as the tissues are coloured naturally brown and yellow.

Ovaries of Flowers.—Make transverse sections, which should be as thin as possible, and stain them either in borax carmine or hæmatoxylin, then clear in clove oil, and mount in Canada balsam.

Anthers must be infiltrated with celloidin to keep the pollen in position. Then embed in carrot, place in microtome, and cut transverse sections. Stain in borax carmine, and after having passed through acidulated spirit, wash well in methylated spirit, and dehydrate for about one to two minutes in absolute alcohol; then clear in oil of origanum or bergamot, and mount in Canada balsam.

Flower Buds.—Infiltrate with celloidin as directed in Chapter VIII. on Section-Cutting. Embed the specimen in carrot, and place in the microtome. Cut transverse sections, stain in borax carmine, and pass through acidulated spirit to remove excess of colour; if desired, they may be soaked until the stain is removed from everything except the nuclei. Wash well in methylated spirit, and place in absolute alcohol for from one to two minutes; then clear in oil of origanum or bergamot, and mount in Canada balsam. Great care must be taken that the sections do not remain too long in absolute alcohol; if they should, the celloidin will dissolve, and the sections will fall to pieces.

Pollens.—Place some mature anthers in a large pill-box, and allow them to become perfectly dry. Shake the box well until all the pollen is set free; then remove the anther sacs with a pair of forceps, and place the pollen in a bottle of turpentine; soak for several days to remove all trace of air, then pour off the turpentine; take up a little of the pollen on the point of a penknife, and place it in a few drops of Canada balsam on a cover-glass; stir up with a needle to spread the grains evenly over the cover, and put away to dry. When the balsam has dried, add a few more drops of balsam, take up the cover with a pair of forceps, and mount it on a warm slide. This method of mounting must always be employed for pollens, because, if they are put up in any other way, the balsam only hardens at the edge of the cover, and remains in a more or less fluid state in the centre, with the result that, if the slide were placed on its edge, the specimens would run together in a heap at the lower side of the cover.

Pollens may also be mounted as opaque objects (see p. 215 on Dry Mounts).

Pollens may also be stained various colours by aniline dyes. Place some fresh pollen in methylated spirit, and soak until air and most of the colour is removed.   Then pour off the spirit and add a strong alcoholic solution of some aniline dye of the desired colour ; any will do so long as it is soluble in alcohol.   Soak in the dye for an hour or two, then pour off the stain, just rinse in spirit, pour this away, and add clove oil, and when clear pour off the oil ; take up a little pollen on the point of a knife, and mount in Canada balsam as directed for unstained specimens.

Specimens of pollens are sometimes stained many colours on the same slide.   This is done in the following way : Take some pollen and divide it into equal quantities, each one of which is to be stained in a different dye.   Then when they have been cleared by the clove oil they are all mixed together and mounted in balsam.

Pharmacological Specimens.—Students of pharmacy may desire to make sections of the dried stems, roots, and leaves with which they deal.   Place the dry specimen in water, and soak until it resumes as nearly as possible its natural shape.   Then place in methylated spirit, which must be changed every twenty-four hours for three days to remove all the water.   Then make sections in accordance with instructions given for ordinary botanical specimens.

Powdered Drugs.—Place some of the powder in methylated spirit, and soak for an hour or two ; then pour off the spirit, and add clove oil ; let it stand a little while, then drain off the oil, take up some of the powder on the point of a knife, place it in some Canada balsam on a cover-glass, mix it well up with the balsam, and then proceed to mount it as directed for pollens.

Some specimens may not be suitable for mounting in Canada balsam ; they should then be mounted in glycerine jelly.   Mix the powder with water, and soak until all trace of air is removed ; allow the mixture to settle down, then pour off the water, take up a small quantity of the deposit, and mount in glycerine jelly as directed for starches.   Sometimes neither glycerine jelly nor balsam will suit these specimens ; then mount dry.

Make an opaque cell, place a little patch of gum-water to its centre ; allow this to dry, then moisten the patch of gum with your breath.   Fill the cell with the powder, and let it stand for a minute or two ; then shake out the powder that has not adhered, and apply a cover-glass.

Mounting in Canada Balsam. — Take 3 ounces of dried Canada balsam and dissolve in 3 fluid ounces of benzole. Filter and keep in an outside stoppered bottle.

1. Clean a cover-glass, moisten the surface of a slide with the breath, apply the cover-glass to it, and make sure that it adheres.

2. Place a few drops of balsam on the cover-glass.

3. Take the section out of the turpentine on a lifter, and put it into the balsam on the cover.

4. Put away out of the reach of dust for twelve hours, to allow the benzine to evaporate from the balsam.

5. Warm a slide over a spirit-lamp and apply a drop of balsam to that on the cover-glass; take it up with a pair of forceps, and bring the drop of fluid balsam in contact with the centre of the warmed slide. Ease the cover down carefully, so that no air-bubbles may be enclosed, and press it down with the point of the forceps until the section lies quite flat and the excess of balsam is squeezed out. Allow the slide to cool, and the excess of balsam may then be washed away with some methylated spirit and a soft rag.

CHAPTER XIV

THE PREPARATION OF VEGETABLE TISSUES FOR MOUNTING IN GLYCERINE JELLY, ACETATE OF COPPER SOLUTION, ETC.

Epidermis for Stomata.—Take a leaf, remove the edges with a pair of scissors, and then cut the remainder up into small pieces of about ¼ inch square. Place these in a test-tube, add nitric acid, and boil gently over a spirit-lamp for about a minute, then add a few grains of chlorate of potash, and bring to the boiling-point again. Pour away the acid and add water, which must be changed several times until all trace of acid is removed. The epidermis will then be found quite clean, and it may be stained and mounted at once, or be placed in spirit and kept until required.

Another Way.—Some epidermal tissues are very delicate and will not stand the acid treatment. When this is the case cut the leaf up as directed above, place the pieces in a jar of water, and put aside for a week or two. The action of water will rot the cellular tissue and set the epidermis free. Then wash well in water, and should any particles of débris adhere, they can be removed by brushing with a camel's-hair brush.

The epidermis of some plants will not stand either of the above processes. When this is the case, the only plan is to strip off a small piece of the cuticle, lay it on a slide, inner side uppermost, and with a scalpel carefully scrape away cellular tissue that may be adhering. Then wash in water, and proceed with the staining.

To Stain the Epidermis.—Make a 1 per cent. solution of methyl aniline violet in distilled water and immerse the specimen for about five minutes. Then wash in a ½ per cent. solution of glacial acetic acid to remove excess of colour, wash away all trace of acid with water, and mount in glycerine jelly in the following way :

Warm the jelly carefully in a water-bath until it is quite fluid. Warm a slide, take up a little jelly in a dipping-tube, and place it on the slide ; now take up the epidermis with a lifter and put into the jelly on the slide, being very careful to avoid making any air-bubbles. Now take the cover-glass and apply it to the surface of the jelly, push down the cover with the point of a needle until it is quite flat, and then set aside to cool. The above process applies to all specimens that are to be mounted in jelly ; but when tissues have been preserved in spirit they must be soaked well in water before being mounted.

Annular Vessels.—Get some stem of maize, cut it into pieces about ½ inch long, and then cut again into thin longitudinal slices ; place these in water until rotten. Now put some of the broken-up material on a slide and examine with a microscope ; pick out the annular vessels on the point of a needle, place them in some clean water, and wash well. Stain in a weak watery solution of acid green, and after washing in water, mount in glycerine jelly.

Scalariform Vessels.—Treat pieces of the rhizome of *Pteris aquilina* in exactly the same way as stem of maize.

Spiral Vessels.—Treat pieces of the stem of rhubarb in the same manner as annular vessels.

Raphides may be isolated, or they can be mounted *in situ*, in the tissues in which they occur. For the former, take some leaves of cactus, stem of rhubarb, and root of Turkey rhubarb, cut them up into thin slices longitudinally, and place them in a jar of water, covered up to keep out dust, and put away until the tissue has become perfectly disintegrated. This will take several weeks, and the process is more easily carried out by keeping the jar in a warm place. When all the material has broken up, stir well with a glass rod, and strain through a piece of coarse muslin into a shallow vessel, such as a soup plate ; stir up again, and then allow to settle for a minute, so that the raphides may fall to the bottom of the plate ; now pour away as much of the dirty water as possible, add more clean water, and repeat the process until you have got rid of all the disintegrated vegetable fibre. Now pour the raphides into a bottle, and if they are quite clean, pour off the water and add methylated spirit, in which they may be preserved until required for mounting.

To mount isolated raphides, clean a cover-glass, fasten it to a slide with the aid of your breath, take up some of the raphides in a dipping-tube, place them on the cover-glass, and spread them evenly over its surface with a needle. Place the slide out of reach of dust until all the spirit has evaporated, and the raphides are quite dry; add a few drops of Canada balsam, and put the slide away again for twelve hours; then add a few drops more balsam, take up the cover with a pair of forceps, and mount it on a warmed slip. When the raphides are very large they must be mounted in balsam that is rather thicker than is usually used.

Raphides in situ in Tissues.—Harden the stems, roots, or leaves in methylated spirit, and make sections in the ordinary way; dehydrate, clear in clove oil, and mount in Canada balsam.

Raphides in Scale-Leaves of Bulbs, such as Onion, Garlic, Lily, Hyacinth.—Strip off a thin portion of the cuticle, place it in methylated spirit for a few hours, and when dehydrated clear in clove oil and mount in Canada balsam.

Sometimes raphides are rendered too transparent when mounted in balsam. When this is the case they must be put up in glycerine jelly in the following way ·

Isolated Specimens.—Pour off the methylated spirit, and add water; pour off the water, leaving the raphides at the bottom of the bottle. Clean a cover-glass and a slide. Place a few drops of warmed glycerine jelly on the centre of the slide; take up a few of the raphides on the point of a penknife, and place them in the glycerine jelly, but do not stir them up. Now apply the cover-glass, and press it down carefully with a needle, giving it at the same time a twisting motion, to spread the raphides evenly between the cover and slide. Put away for an hour or two, scrape off the excess of jelly with a penknife, wash in water, and then in methylated spirit, dry with a cloth, and apply a coat of black enamel. When raphides in the tissues are prepared in glycerine jelly, wash away all trace of spirit with water, and mount in glycerine jelly as above.

Starches (*Isolated Specimens*).—If the tissue is fresh, scrape the cut surface with a knife, and place the scrapings in a bottle of water ; shake well and then strain through fine muslin into a shallow vessel; let the starch settle, pour off the water, and wash again with some clean water until the starch is quite clean; then place it in a bottle, and when it has settled to the bottom, pour off the water, and add methylated spirit.

Dried Specimens.—Place in water until the tissue swells up, then, if the material is large enough, it may be scraped and treated as above. If too small—small seeds, for instance—place them in a mortar in some water, and carefully break them up ; strain through muslin, wash with water until quite clean, and preserve in methylated spirit.

Starches may be mounted in Canada balsam or glycerine jelly. If the former is chosen, spread a little starch evenly on a cover-glass, let it dry, apply some Canada balsam, and mount it in the ordinary way. For glycerine jelly pour off the spirit and add water, then allow the starch to settle to the bottom of the bottle; pour away the water. Place a few drops of glycerine jelly on a slide, take up some starch on a penknife, and place it in a little heap in the jelly ; now apply a cover-glass, and press down with a gentle twisting movement until the starch is evenly spread. Let the jelly set, scrape away the excess, wash in water, then in spirit, dry, and apply a coat of cement.

It is desirable also to prepare specimens of starch *in situ* in the tissues. Take, for example, a potato, cut it into small pieces of about ½ inch square, and harden them in methylated spirit. Then embed in carrot and cut the sections, which should not be too thin. Stain in a 1 per cent. solution of methyl aniline violet, wash in water, and mount in glycerine jelly.

In mounting starch in glycerine jelly, care should be taken that the jelly is not too hot ; if it be, the form of the starch will be altered.

Yeast.—Get some fresh baker's yeast, place a little of it in a bottle of sugar and water, and stand in a warm place for twenty-four hours. Pour off the sugar water, and add camphor water. Make a cell on a slide with black shellac cement, and let it dry ; then apply a second coat of cement, and let this stand for a few minutes. Now take up some of the yeast in a glass tube and place a few drops in the cell; clean a cover-glass, and bring its edge in contact with the cement on one side of the cell; ease it down carefully, so that no air-bubbles may be enclosed; now press on the surface of the cover with a needle until it adheres firmly to the cell all round, drain off the excess of fluid, dry the slide with a clean cloth, and apply a coat of cement.

Mycetozoa or Myxomycetes.—Most of these fungi can be mounted in glycerine jelly after soaking in equal parts of rectified spirit and glycerine to remove the air, but in those forms which possess lime granules in the capillitium—a character of importance in classification—the calcareous matter disappears when in glycerine in any form. When this is the case, place the specimen in absolute alcohol until all air is removed, then transfer to clove oil, and mount in Canada balsam. Some specimens may, however, be rendered too transparent by the balsam; if so, mount them in a shallow cell in some neutral fluid such as camphor water.

In their ripe condition they may also be mounted dry as opaque objects.

Large fungi, such as Agaricus, should be hardened in methylated spirit for a week. Then place the desired portion in water, and soak to remove spirit, transfer to gum and syrup, and when penetrated with the gum, freeze and make the sections with a Cathcart microtome, wash away all trace of gum with repeated changes of warm water, and mount unstained in glycerine jelly.

Preserving Fluid for Green Algæ.—Acetate of copper, 15 grains; camphor water, 8 ounces; glacial acetic acid, 20 drops; glycerine, 8 ounces; corrosive sublimate, 1 grain. Mix well together, filter, and keep in a stoppered bottle. The above fluid preserves the colour of chlorophyll for a long time; it may also be used as a mounting fluid. For very delicate specimens leave out the glycerine.

The specimens should be well washed in water; then pour off the water, and add a quantity of the copper solution.

To Mount in the Above.—For example, take Spirogyra as a filamentous alga. Make a cell with some black cement, and let it dry; then apply a second coat of cement, and allow this to nearly dry. Place some Spirogyra in the cell, and with needles separate the filaments ; add a few drops of copper solution, and apply a cover-glass as directed for yeast.

Protococcus.—This can be obtained by scraping the bark of trees. Place it in a bottle of water, and let it stand for a few hours : now add a little copper solution—this will kill the specimens, and they will sink to the bottom of the bottle ; pour off the water, and add more copper solution. Now make a cell as for spirogyra ; take up some of the protococcus in a dipping-tube, and place them in a cell ; wait a minute for the forms to settle on the bottom of the cell, and then apply a cover-glass ; drain off the excess of fluid, dry the slides with a cloth, and apply a coat of cement.

Volvox, glœocapsa, desmids, etc., may all be preserved and mounted as above.

Antheridia and Archegonia of Mosses.—Place some male and female heads of mosses in methylated spirit for a few days, then transfer to equal parts of absolute alcohol and ether, in which they must be soaked for several hours. Pour off the alcohol and ether, and add a thin solution of celloidin, and soak for two or three days ; then remove the stopper of the bottle, and let the celloidin evaporate to about half its original bulk. Now remove a specimen from the celloidin, and hold it in a pair of forceps until the celloidin sets, then place it in methylated spirit and soak for an hour or two to complete the hardening. The embedded specimen may now be fastened to a cork with a little celloidin, and longitudinal sections made in a Cathcart microtome, or it can be placed between two pieces of carrot, and the sections made with any ordinary well microtome. The sections must then be dehydrated in methylated spirit, cleared in oil of bergamot, and mounted in Canada balsam ; or, if desired, they may be soaked in water to remove spirit, and be mounted in glycerine jelly.

Fertile Branch of Chara.—Chara is usually very dirty ; to clean it, wash well in repeated changes of water, then in very dilute acid for a few minutes only ; again wash in water, and preserve in camphor water.

Make a cell with shellac cement as directed above, place a fertile branch of Chara in it ; and examine under a dissecting microscope or lens ; with needles clear away the leaves from the archegonia and antheridia, fill the cell with camphor water, and apply a cover-glass.

When a deep cell is required for a specimen to be mounted in acetate of copper, never use one made of any metal. Vulcanite or glass cells must be used. To one side of a cell apply a coat of shellac cement and let it dry ; now take a slide and warm it over a spirit-lamp ; take up the cell in a pair of forceps, and bring the cemented side in contact with the centre of the warm slide, and press it down until it adheres firmly ; then add another coat of cement to the upper side of the cell, and let it nearly dry, put in the specimen, fill the cell with solution, and apply the cover-glass.

Prothallus of Fern.—Preserve in acetate of copper and mount in the same fluid in a shallow cell.

Sporangia and Spores of Fern.—Place leaves of a fern with sporangia in methylated spirit for a few days to remove the air. Then soak in water for several hours. Warm a slide, and place a few drops of glycerine jelly on its surface, scrape off some sporangia, and place them in the jelly ; now apply the cover-glass very carefully to avoid scattering the sporangia. The object is to keep them in a heap in the centre until the cover is flat ; then press on the surface of the cover with the points of the forceps, and, if possible, give the cover a little twisting motion. This will spread the specimens ; it will also rupture some of the sporangia and let out the spores.

Isolating Antheridia and Oogonia from Fucus.—Take some conceptacles that have been hardened in methylated spirit, and make thick sections by hand only with a sharp knife. Place these in a strong solution of acid aniline green in spirit, and let them stand for two or three hours. Now place in water for a few minutes, and they will at once swell up like a mass of mucus. Place this on a slide and put another slide on top of it, press down the upper slide—this will squeeze out the contents of the conceptacles in little round masses. Separate the glasses, pick up one of the little lumps of antheridia or oogonia, place it in a few drops of glycerine jelly on a slide, then apply the cover-glass, which must be pressed down to spread the specimens.

Digestive Glands in Pitcher Plant.—Harden some strips of a pitcher in methylated spirit for a week. Then place in water and soak for a few hours. Then lay the tissue with the glandular surface next to the glass, and with a scalpel scrape away the outer wall. Now bleach the glandular portion in chlorinated soda, then wash well with water, stain in aqueous solution of acid aniline green, wash again in water to remove excess of colour, soak for several hours in dilute glycerine, and mount in glycerine jelly.

Aleurone.—Take the endosperm of a castor-oil seed, embed in carrot, place in microtome, and cut sections as thin as possible with a knife wetted with a little olive oil. As the sections are cut, put them on a slide, and place out of reach of dust until you are ready to mount them.

Make a shallow cell as directed with black enamel and let it dry, then proceed as directed for acetate of copper mounting, but use castor oil instead of copper solution. When the cover has become fixed, wash away the exuded oil with a soft brush and some turpentine, and, when dry, apply a good finishing coat of black enamel. Water and spirit are apt to injure the aleurone grains, so they should be avoided.

Marine Algæ.—The best place for collecting specimens is a rocky shore, and the most suitable time is when the tide is at its lowest. As a rule, the inshore weeds near high-water mark are green, lower down there is usually a belt of olive forms sheltering red plants beneath them, and where rocks overhang small shallow pools red forms also occur at this level. At extreme low-water mark and beyond it are found brown tangles sheltering red plants again, while at the lowest depths the red weeds occur without shelter. The specimens will be found by searching the rocks and pools, some will be growing on pebbles and on shells, others will be attached to rocks, and varieties may be found stranded on the shore, thrown there by waves, particularly after a storm, the tufts having been torn away and carried inshore from inaccessible regions.

For collecting, small tin boxes or an ordinary sponge bag will be found most suitable. A strong chisel mounted on a stout stick will also be required for removing specimens from rocks that are out of reach.

Many specimens may be preserved in sea water for a considerable time, but, as a rule, the sooner they are mounted the better.

Mounting Process.—Remove the specimen from sea water and wash well in fresh water. Place in a shallow white dish or saucer, select and cut off the portion that is to be mounted, and place it on a slide slightly warmed, drain away as much water as possible, and apply some glycerine jelly ; then, if necessary, lay or spread out the leaves or filaments with a needle and apply the cover-glass, allow the slide to cool, remove the excess of jelly around the edge of the cover, wash the slide in water, dry, and add several coats of enamel or varnish.

Corallines, whose tissues are hard and opaque, may be cleaned by soaking for a short time in a weak solution of hydrochloric acid, then wash well in water, and mount in glycerine jelly.

CHAPTER XV

CUTTING, GRINDING, AND MOUNTING SECTIONS OF HARD TISSUES—PREPARING METAL SPECIMENS

Bone.—Take the femur of a sheep, remove as much of the muscle as possible, and macerate in water until quite clean, then allow it to dry.

1. With a fine saw make transverse and longitudinal sections.

2. Take a hone (water of Ayr stone), moisten it with water, and rub one side of the section upon it until it is quite flat and smooth.

3. Wash in water, and set aside until quite dry.

4. Take some dried Canada balsam, place a piece on a square glass, and warm gently over a lamp until the balsam melts; allow it to cool a little, and then press the smooth side of the section into it, and set aside until cold.

5. With a fine file rub the section down as thinly as possible.

6. Take the hone again and grind the section down until thin enough, using plenty of water.

7. Place it, with the glass, in methylated spirit until the section comes away from the glass, then wash well in clean water and allow to dry.

8. Place the section on a slide, and apply a very thin coat of gum water to its upper surface, taking great care that the gum does not run under the section, and let it dry. This coat of gum will hide any fine scratches that may be left on the section. Now take a thin cell just deep enough for the section, and apply a coat of cement to its upper edge; place the section in its centre with the gummed side uppermost, and apply the cover-glass, which should come down on the bone to keep it in the centre, hence the necessity of a cell of just the proper depth. The object of a section of dry bone is to show the canaliculi;

194

when mounted in fluid of any kind these are obliterated. Sections of teeth are made in the same way.

Rock Sections.—Small pieces or slices of rock are to be ground on a zinc plate with the aid of emery powder and water until one side is quite flat and smooth. Then fasten the polished surface to a square of glass with some dried Canada balsam, as directed for bone, and allow it to cool. Grind the other side on the zinc plate with coarse emery and plenty of water. When moderately thin, take a piece of plate-glass and some fine flour-emery, and rub the section down as thinly as possible. When thin enough, wash well in water and dry ; then warm over a spirit-lamp, and with a needle push the section off the glass into a saucer of benzole or turpentine, and allow it to soak until all the balsam is dissolved. Wash again in some clean benzole, and mount in Canada balsam in the usual way. Sections of echinus spines, shells, and stones of fruit are prepared in the same way as bones and teeth ; but when the grinding is finished, the sections are to be passed through alcohol into clove oil, then mount in Canada balsam in the usual way.

Sections of coal containing fossils, limestone, spines of echinus, and other friable specimens should be cut with a very fine saw, and then soaked in benzole for several hours. When the benzole has saturated the tissue, transfer to ordinary solution of Canada balsam in benzole, and soak again until the balsam has penetrated to the centre. Take a 3×1 inch slide, place the section on its centre, and add sufficient balsam to cover it. Put away out of reach of dust until the benzole has evaporated from the balsam. Then place on a hot plate, apply gentle heat with a spirit-lamp, and bake until the balsam is quite hard. Grind down to the required thinness on a hone. Wash well with water, dry, add a few drops of fluid balsam in benzole, and apply a cover-glass.

Metal Specimens.

The preparation of specimens of metal for the microscope involves the greatest care, the principal object being to obtain a perfectly level surface, free from all scratches and marks, with the highest degree of polish. This will be better illustrated by an example.

The student having obtained a sample of metal, the first thing to do is to carefully file or grind the surfaces he wishes to examine. The marks thus made must be taken out with a very smooth file or emery cloth, gradually diminishing the coarseness of the cloth until he reaches the finest grade of all.

From this stage the polishing must be done on parchment or chamois leather stretched very tightly on wood, the leather being covered with fine crocus powder or rouge moistened with a little water.

This is the most important stage of the specimen, especially if the metal be very soft, and the student should frequently examine the metal through the microscope—a matter of a few moments only—by clamping it in the new metal-holder recently introduced by Messrs. Watson, as shown on p. 41.

It will then be seen that parts stand in very high relief. The object of the leather polishing being to gradually grind away the soft and leave the hard parts, great care must be exercised in doing this.

The specimen is now ready for further treatment—viz., **etching**. The object of this is to further develop the structure, as will be seen from below.

Etching.—This is done by various reagents, the choice of which is mainly a matter of personal opinion, but perhaps the most generally used, and the best for beginners, are infusion of liquorice root and tincture of iodine. Very dilute nitric acid and sulphuric acid are also used, but until the student has become thoroughly acquainted with the effects of the above he is not advised to use them.

Before proceeding further it is advisable to give an outline respecting the effects of the reagents; also the construction of the metal.

Steel is viewed as if it were a rock with various constituents in it. There are three principal ones—viz., ferrite, cementite, and pearlite (or sorbite).

Ferrite.—This is iron free from carbon. It retains a very dull polish, and is not stained by iodine or liquorice.

To develop the crystalline structure of ferrite a very dilute solution of nitric acid in alcohol should be used.

Cementite.—This is a very hard substance, and stands in relief after polishing, as above. It is very rarely found in low carbon steels, and is left bright after the polished surface is attacked by iodine.

Pearlite.—This is a very intimate mixture of ferrite and cementite. If the steel has been allowed to cool slowly from a very high temperature, pearlite assumes a well-defined lamellar structure ; on the contrary, if the metal has been forged or reheated at a very low temperature, pearlite assumes a granular appearance. It is readily acted upon by iodine or liquorice.

From this it will be seen that steel is made up of (1) ferrite and pearlite, (2) pearlite, (3) pearlite and cementite. Other constituents are found in steel after it has undergone certain treatment, but enough has been said to guide the student to make a commencement.

The reagent can now be applied. This is done by either coating the specimen with some protective varnish, leaving the surface free that is to be acted upon, and immersing the whole in a bath, or a few drops may be applied to the surface, and then carefully spread by means of a glass dipping-rod.

The solution should be allowed to act for, say, twenty seconds, then carefully washed in alcohol or methylated spirit, gently rubbing the surface with the little finger, finally washing in water and drying with a very soft piece of linen.

The metal is now examined under the microscope, and it will then be seen if the etching has been sufficient ; if not, it should be repeated, as above, for another twenty seconds. The student should do this several times, noting the effect of the reagent each time until he becomes thoroughly acquainted with its properties.

So far we have only dealt with steel, but alloys of tin, copper, etc., are treated in exactly the same way, with the exception that liquorice and iodine are not used. The various acids, ammonia, and caustic potash are then used in weak solution as etching reagents.

With respect to the mounting of the specimens, it will be seen that the new holder does away entirely with the glass slide. It often happens that the lower edge of the metal is left jagged, and may also be broken off at a very sharp angle, necessitating a long delay in filing or grinding ; also the metal must not be too thick if a glass slide is used ; but, as will be seen, this labour is greatly minimized if the holder be used.

CHAPTER XVI

PREPARING AND MOUNTING ENTOMOLOGICAL SPECIMENS FOR THE MICROSCOPE

INSECTS should be killed with chloroform. They are then to be placed in methylated spirit, in which they may remain until required for mounting.

To Prepare a Whole Insect for Mounting with Pressure in Canada Balsam.—1. Transfer from methylated spirit to water, and let it soak for three or four hours to remove spirit.

2. Place in liq. potassæ—10 per cent. of caustic potash in distilled water—until soft. Some specimens will only require a few hours in the potash, others need days, and some even weeks, to soften. In all cases they must be carefully watched and the action of the potash tested. This can be ascertained by pressing on the thorax or chest of the insect with some blunt instrument such as the head of a pair of curved-pointed forceps.

3. When soft enough, pour away the potash and add water, which must be changed several times until all the potash is washed away.

4. Pour away the water and add concentrated acetic acid, and soak for twelve hours, or until it is convenient to go on with the work.

5. Transfer from acetic acid to water, and soak for about half an hour ; then place in a shallow saucer full of water, and with the aid of a needle and a camel's-hair brush spread out the wings, legs, etc. Now take a slide and place it in the water under the insect, lift the slide up carefully so that the insect may be stranded on the surface of the slide with all its parts expanded. Drain off the excess of water, and lay the slide down on a piece of white paper, and with the aid of needles or brushes

carefully place all the limbs, wings, antennæ, etc., in their natural positions.   Now put a narrow slip of paper on each side of the insect, and carefully lay another slide over it, press it down until the insect is squeezed quite flat, tie the two slides together with a piece of twine, and place them in a jar of methylated spirit for at least twelve hours, or until required.

6. Remove the glasses from the spirit, carefully separate them, and with a soft camel's-hair brush push the insect off the glass into a saucer of spirit, and soak for half an hour.

7. Take the insect up on a lifter, and float it on to the surface of a small saucer of clove oil, and allow it to soak until perfectly clear.

8. Remove from clove oil and place in turpentine for a few minutes.

9. Mount in Canada balsam as directed for animal and botanical sections.

To Mount an Insect in Canada Balsam without Pressure. —Treat with potash as above, wash in water, and place in acetic acid.   Wash away the acid with water, and transfer to a shallow saucer of methylated spirit.   Take two needles and lay out the various parts as quickly as possible ; if any parts are troublesome, hold them in position until the spirit has fixed them.   Now let it soak for an hour, or until required.   Remove from spirit, place in clove oil, and when clear, place in turpentine.

Take a *tin* cell just deep enough for the specimen, and apply a coat of black shellac cement to one side of it.   Allow this to nearly dry.   Clean and warm a slide over a spirit-lamp ; take up the cell in a pair of forceps, and bring the cemented side in contact with the centre of the warmed slide ; press on the upper side of the cell, until it adheres firmly to the slide, and put it away to dry.   Fill the cell with Canada balsam, and see that it also flows over the upper edge of the cell, so that it may serve as a cement to fasten on the cover.   Take the insect from the turpentine on a lifter, put it in the cell, and with needles rearrange the parts if necessary.   Put away out of reach of dust for twelve hours to harden the balsam.   Place a drop of balsam on one side of the cell.   Clean a cover-glass of the same size as the cell, take it up in a pair of forceps, and warm it gently over a spirit-lamp, and bring its edge in contact with the drop of fresh balsam ; ease down carefully, so as to avoid air-bubbles, and press on surface of cover with a needle until it rests on the cell all round. Now take a soft brush and some benzole and wash away the exuded balsam ; dry with a clean rag, and apply a ring of cement.

To Mount an Insect in Glycerine without Pressure.— Many small, soft insects and their larvæ may be mounted in glycerine while fresh. The larger and harder kinds must be soaked in potash to render them transparent. Make a cell of the required size, and fasten it to a slide with black shellac cement, as directed for balsam mounts. Apply a coat of cement to the upper side of the cell, and allow it to nearly dry. Fill the cell with glycerine, and put the insect into it; spread out the wings, legs, etc. Clean and warm a cover-glass, and apply its edge to the cell ; press down, and be sure that it adheres to the cement all round. Wash away the excess of glycerine with some water, and dry the slide with a soft cloth. When quite dry, apply a ring of cement, and when this has dried, add another coat of black shellac cement.

The processes described only refer to the study of the external parts of insects ; all the soft tissues and internal organs will, of course, have been destroyed by the potash. Soft internal organs must be dissected out of the specimen while under water.

Procure a guttapercha dissecting-dish, lay the insect in it, and secure with pins in the desired position. If the abdominal or thoracic viscera are required, lay the insect on its back; if the nervous system, on its ventral surface. Fill the dish with water, and with a pair of sharp-pointed scissors cut through the chitinous skin on each side of the abdomen, taking care not to cut too deeply so as to injure the internal organs; then with a pair of forceps raise and remove the skin. The organs may now be removed with the aid of a pocket-lens, and washed in distilled water ; then stain in borax carmine for several minutes, wash in methylated spirit; then immerse in acidulated alcohol for a few minutes, dehydrate, clear in clove oil, and mount in Canada balsam.

If desirable to mount the specimen in glycerine, stain as above, then wash away all trace of spirit with water, and mount in glycerine jelly; if the specimen requires a cell, it must be mounted in glycerine.

Salivary glands of cockroaches and crickets, gizzards of beetles, and stings of bees and wasps, may be easily removed in the following way: Place the specimen whole and while quite fresh in water, cover with a piece of paper or anything to keep out dust, and let them soak for several days until the smell becomes rather unpleasant; then wash in clean water, hold the insect between the fingers, and with a pair of forceps carefully pull off the head, which should bring with it the œsophagus, salivary glands, and stomach. For stings of wasps and bees proceed as follows : Gently squeeze the abdomen of the specimen between the fingers of the left hand until the sting protrudes, then grip it with a pair of fine forceps, and gently pull it out. If properly done, the poison gland and duct should come away with it. Wash in water, and place it on a slide under a dissecting microscope, and with a fine needle-point draw the stings from their sheath; this is done by putting the needle under the stings at the base of the sheath and carefully drawing it towards the apex. Stain in borax carmine, wash in alcohol, then in acidulated alcohol, and place in water; now lay out on a slide, place another slide over it, tie with thread, and immerse in methylated spirit for several hours ; remove from glass, clear in clove oil, and mount in Canada balsam.

Small insects, such as parasites, may be mounted whole in a cell in glycerine without treatment with potash, so that their internal organs may be seen *in situ*, but they usually require clearing. Take of Calvert's carbolic acid, solid at ordinary temperatures, 2 ounces, melt, and add about ½ a drachm of glycerine to prevent it becoming solid again. Soak the insect in this until transparent ; some specimens will only require an hour or two, others a week or more. When clear, make a cell as previously directed with any good shellac cement, and when dry, run on a coat of cement to its upper surface; let this become about half dry, then place in the cell, fill it up with glycerine, and apply a cover-glass, which must be carefully pressed down with a needle-point until it adheres to the cement all round. The slide can then be washed with water to remove all trace of excess of glycerine ; put away until all the water has evaporated, then apply a coat of shellac cement, and when this has dried, rub away any water-marks that may be left on the slide with a soft cloth, and add another coat of cement.

Wing-cases, legs, heads, and feet of diamond beetles should be mounted in opaque cells in Canada balsam. Take a slide, and with a turn-table run on a disc of black varnish of the required size; allow this to dry thoroughly. Take a piece of black gummed paper and punch out a disc of the same size as that on the slide to which it is to be fastened. Now take a *tin* cell of the required depth—on no account use brass or vulcanite cells; they are affected by the balsam, and the mount will be spoiled—lay the cell on a slide, and apply a coat of cement to its upper surface; allow this to become nearly dry, then take up the cell in a pair of forceps, and bring its cemented surface in contact with the paper disc on the slide, and with the point of the forceps press the cell down until the cement adheres to the paper. Now put away to dry in some place protected from dust. Take the specimen to be mounted, examine it under a microscope, and if dirty wash in some benzole, and then let it dry again. Now place a small quantity of gum-water in the centre of the cell, and put the specimen into it in the desired position; make sure that it adheres securely to the gum, and put the slide away again until everything is quite dry. Put the slide in a turn-table, and run on a coat of shellac cement to its upper surface, and allow it to become nearly dry; then fill up the cell with Canada balsam, clean, and apply a cover-glass, which must be well pressed into the cement until it adheres firmly; put away for an hour, and then wash away the exuded balsam with a soft brush and some turpentine; dry the slide with a soft rag, and apply a coat of black shellac cement.

Heads of flies having coloured compound eyes, such as Tabanus, lace-wing flies, etc., should be mounted in opaque cells in glycerine. Make the cell in exactly the same way as directed for balsam mounts, but take care that the cell is only just deep enough to take the specimen, as the object has to be retained in the centre of the cell by slight pressure on the part of the cover-glass. When the cell is quite dry, apply a coat of shellac cement to its upper surface, and let it nearly dry; then take a brush and some clean water and moisten the inside of the cell. This is done to prevent the formation of air-bubbles, for if glycerine is put into a dry cell, bubbles are sure to give a lot of trouble. Now fill the cell with glycerine and put in the specimen, which should be previously soaked in dilute glycerine for an hour or two, and with a needle place it in the desired position ; apply the cover-glass very carefully, so that no air-bubbles may be enclosed, and let it settle down by its own weight until it rests on the surface of the cell ; then press it down with a needle-point until securely embedded in the half-dried cement, and set aside for an hour or two to dry. The exuded glycerine may then be washed away by holding the slide under a water-tap. When all trace of glycerine is removed, dry the slide with a soft cloth, and apply a coat of black shellac enamel.

Heads of large insects may be secured in the centre of the cell in the following way : Take a fine needle, thread it with a hair, and run it through the specimen. Unthread the needle, take up each end of the hair with the object suspended and stretch it across the cell so that it may be embedded in the cement on each side. Now apply a cover-glass, press it down until securely fixed, and if the specimen is not in the middle of the cell, adjust it by pulling on the hair on one side. Put away to dry, cut off the ends of the hair close to the edge of the cell, wash away excess of glycerine, dry, and apply a coat of shellac enamel.

CHAPTER XVII

CRYSTALS AND POLARISCOPE OBJECTS

Crystals.—*Method* 1.—Make a strong solution of the material in distilled water, with the aid of heat if necessary, and filter; take up a small quantity of the solution in a dipping-tube, and drop it on a cover-glass. Prepare several covers in this way, and allow some to dry slowly, and evaporate others over a spirit-lamp. When dry, add a drop or two of Canada balsam, and mount in the usual way.

Method 2.—Make a strong solution in distilled water, and add a few drops of gum water or a small piece of gelatine; mix well, and filter. Apply some of the solution to a cover-glass, and allow it to dry slowly in a place protected from dust. Mount in Canada balsam.

Method 3.—Place a small piece of the dry crystal on a slide, and apply a cover-glass; warm over a spirit-lamp until fusion results, press the cover down with a needle, and allow the slide to cool. Clean off the exuded material, and finish off with some good cement.

Some crystals are soluble in Canada balsam; in which case, mount in castor oil.

Crystallize the specimen on the cover-glass; make a thin cell with some shellac cement on a slide, and allow it to become perfectly dry; then apply another coat of cement, and when this has nearly dried, fill the cell with castor oil. Take up the cover with a pair of forceps, and bring the crystallized surface in contact with the oil, being very careful that no air-bubbles form. Ease it down gently, and when it rests on the cell, give it a press with the point of the forceps; this will squeeze out the excess of oil and embed the edge of the cover in the cement.

204

Put away to dry ; wash off the exuded oil with some turpentine, and apply another coat of shellac cement.

The following salts, etc., are easily obtained, and they all give very good results

Chloride of barium.*	Sulphate of iron.*	Asparagine.
Chlorate of potash.*	Tartrate of soda.*	Quinidine.
Sulphate of copper.*	Salicine.	Santonine.
Spermaceti (fuse).	Stearine (fuse).	Tartaric acid.

Those marked * are more effective when crystallized in gum or gelatine.

Crystals of Silver.—Clean a cover-glass and fasten it to a slide with the breath ; make a 1 per cent. solution of nitrate of silver, and place a drop of it in the centre of the cover-glass. Now add a very small fragment of copper, and put the slide away out of reach of dust until the crystals have formed and all moisture has evaporated. Then make a shallow opaque cell, and place a small drop of gum water in its centre. Take up the cover with a pair of forceps—crystals uppermost, of course—and drop it into the cell ; now take a needle-point, and carefully press on the cover-glass between the crystals, until it lies quite flat, and air-bubbles, if any, have exuded. Put the slide away again until the gum has dried. Now put the slide into a turn-table ; run on a coat of shellac cement to the upper surface of the cell. Allow this to become half dry, and then apply a cover-glass.

The following specimens from the vegetable kingdom make fine polariscope objects : Starches, hairs, scales from leaves, cotton and silk fibres, cuticles of leaves, and longitudinal and transverse sections of stems.

Starches can be obtained from most vegetable substances by scraping the cut surface with a knife. Place the scrapings in a bottle of water and shake well ; then strain through muslin of sufficiently fine texture to allow the starch to pass, but to retain the fibres. Now put the strained material into a bottle, shake it up, and then allow to settle ; the starch will fall to the bottom of the bottle in a few minutes. Then pour off the water ; add some more, and repeat the process until all trace of cellular tissue is removed. When the starch is quite clean, take up a little in a dipping-tube; apply a drop to a clean cover. See that it spreads evenly all over the surface of the cover, and put away, protected from dust, until quite dry; then add a drop of Canada balsam, and mount in the ordinary way.

Starches may also be mounted in glycerine jelly (see p. 188), but they do not polarize so well as the balsam preparations.

Sections of Starch-Bearing Tissues.—The stems, roots, and bulbs must be hardened in methylated spirit for a week; then make transverse or longitudinal sections. Dehydrate in methylated spirit, clear in clove oil, and mount in Canada balsam.

Cuticles containing Raphides.—The most common are taken from the following bulbs : garlic, onion, lily, hyacinth. Strip off the cuticle from the fresh specimen; dehydrate in methylated spirit, clear in clove oil, and mount in Canada balsam.

Cuticles of Leaves.—Cut up the leaf into small pieces, and soak in water until rotten; the cuticles can then be separated, washed in water, dehydrated in methylated spirit, cleared in clove oil, and mounted in Canada balsam.

Cotton, Hemp, Wool, Silk, Flax, etc.—Place the fibres in methylated spirit to dehydrate; then clear in clove oil, and place a little on a slide. Separate the fibres from each other with needle-points; apply a few drops of Canada balsam and a cover-glass.

Scales of Leaves.—Scrape the leaf with a knife, and put the scrapings into a bottle of turpentine, and soak until all trace of air has disappeared from the scales; then pour off the turpentine. Take up a little of the scales on the point of a penknife, and mount them in Canada balsam in the ordinary way. Some leaf-scales are very difficult to deprive of air; in fact, it is impossible to get them quite free.

The following animal tissues make good polariscope objects : fish-scales, palates of molluscæ, sections of hairs and quills, horns and hoofs, whalebone, claws of dogs, cats, and fowls, decalcified bones, muscular tissues.

Fish Scales.—Scrape the fish from the head towards the tail; if scraped the other way, nearly all the scales will be injured. Place the scrapings in a bottle of water, shake well, pour off the water, and repeat the process until quite clean. Examine with a microscope, and if you find that the scales are not clean, pour off the water, add liq. potassæ, and soak for an hour or two; then wash away the potash with repeated changes of water, dehydrate in methylated spirit, clear in clove oil, and mount in Canada balsam.

Sometimes fish scales buckle up in spirit, and they will not lie flat. When this happens, put them into water again, and soak a little while; then place them on a slide, and put another slide over them, press down until quite flat, and tie the two glasses together with twine, and place them in a vessel of methylated spirit to dehydrate under pressure. This method will answer for all tissues that have a tendency to twist during the process of dehydration.

Palates.—Dissect out, and soak in liq. potassæ for a few days. Wash well in water, spread out on a slide; put a piece of paper on each side of it to prevent crushing, and place another slide over all in the same way as directed for insects; tie the glasses together with string, and place in methylated spirit for an hour or two. Then remove the palate from the glasses, and place it in clove oil until clear. Mount in Canada balsam.

Sometimes it is very difficult to dissect out the palates from small snails. This process answers just as well: Cut off the head of the animal, being careful that you remove the buccal mass with it, and place in liq. potassæ for a few days; this will destroy all the soft tissues, but not the palate or radula. Wash away the potash with repeated changes of water, and proceed as directed above.

Sections of Hairs and Quills may sometimes be cut after soaking for a few days in methylated spirit; but some of the larger kinds, such as the whisker of walrus, will require softening in potash. Place in liq. potassæ for a few hours or days, in accordance with the consistency of the tissue. When soft enough, wash away the potash with water, and place in methylated spirit, in which they may be preserved until required. Then make transverse and longitudinal sections, dehydrate in methylated spirit, clear in clove oil, and mount in Canada balsam.

Small Fine Hairs.—Cut off a number of hairs, tie them up into a bundle with some cotton, and soak for a few minutes in warm water. Make up a strong solution of gelatine in water, and transfer the bundle of hairs to it, and soak it for several hours in a hot-water bath until the gelatine has penetrated to the centre of the bundle. Remove from the gelatine on the point of a needle, and hold it exposed to the air until the gelatine has cooled ; then push them from off the needle into a bottle of methylated spirit, and soak for an hour or two to complete the hardening. Embed in carrot, put in a microtome, and cut transverse sections, and as they are cut place them in methylated spirit to dehydrate ; then clear in clove oil, and mount in Canada balsam.

Horns, Hoofs, Whalebone, and Claws all require steeping in liq. potassæ until soft ; they are then to be washed in water, and preserved in methylated spirit until required. Embed in carrot, place in a well microtome, make transverse and longitudinal sections, dehydrate in methylated spirit, clear in clove oil, and mount in Canada balsam.

Decalcified Bones (see Chapter VII.).—Embed in carrot, make transverse and longitudinal sections, dehydrate in methylated spirit, clear in clove oil, and mount in Canada balsam.

Muscular Fibres.—Take the tongue of a cat, harden it in methylated spirit for a week or ten days ; then embed in carrot, and make transverse or longitudinal sections, dehydrate, clear in clove oil, and mount in Canada balsam.

CHAPTER XVIII

CLEANING AND MOUNTING DIATOMS, POLYCYSTINA, AND FORAMINIFERA

To Clean Diatoms growing upon Algæ or Shells.—Place the algæ or shells in a basin, cover them with water, add hydrochloric acid, and stir until effervescence results ; add more acid little by little, until effervescence ceases, stirring from time to time. Now strain through net of sufficiently fine texture to allow the diatoms to pass, but to retain the débris. Allow the strained fluid to settle down, pour off the acid water, and place the deposit in a large test-tube. Add pure hydrochloric acid, and boil for twenty minutes ; add some pure nitric acid, and boil again for twenty minutes, and, while boiling, add some crystals of chlorate of potash until complete bleaching results. Remove all trace of acid or alkali by washing in water, and examine the forms under the microscope. If clean, bottle them up in distilled water for future mounting. If, as is sometimes the case, there has been animal matter present which has not been removed, boil in pure sulphuric acid for a few minutes. Wash away all trace of acid before bottling the diatoms in distilled water.

To Clean Fossil Diatomaceous Deposits.—Break the deposit up into small pieces, and place them in a large test-tube in a moderately strong solution of bicarbonate of soda, and boil gently for two hours, the disintegrated portions being from time to time poured off into a beaker and the boiling in soda continued until all the deposit has broken up. The alkaline solution must then be washed way, and the diatoms boiled for a short time in nitric acid, and when sufficiently clean wash away the acid in repeated changes of water, and bottle up the diatoms in distilled water.

To Clean Living Diatoms.—Remove all dirt or salt by washing well in water; shake well, and allow the diatoms to settle before pouring off the water. In this way all soluble impurities can be removed. When the water remains clear, pour it off, leaving the diatoms as nearly dry as possible, and cover them with strong alcohol, which will extract the endochrome; change the alcohol daily until it ceases to be tinged with green; then wash away the alcohol with water, pour off the water, and place the diatoms in a platinum capsule and heat them to a dull red over a spirit-lamp. This will separate the frustules into single valves, and finish the cleaning of the diatoms, and they may then be bottled up in distilled water.

To Clean Polycystina.—The polycystinous earth should be broken into small pieces and boiled for several hours in a strong solution of common washing soda, the disintegrated matter being from time to time poured off into a vessel, and the boiling in soda continued until all the earth is broken up. Wash the disintegrated matter in water several times to remove the soda, allow the polycystina to settle down, and pour off the water and place the forms in a test-tube; add some nitric acid, and boil for twenty minutes. Remove all trace of acid with water, and bottle up in distilled water.

To Clean Foraminifera.—All mud must be got rid of by repeated washing in water. Then boil the forms in a strong solution of bicarbonate of soda for an hour or two. When clean, wash away the soda, and bottle in distilled water.

To Mount Diatoms in Canada Balsam (*Unselected Slides*). —The diatoms are to be taken out of the bottle with a dipping-tube, and should be allowed to fall upon a clean cover-glass. The fall of the drop causes the forms to spread evenly over the cover. It should then be dried slowly over a spirit-lamp. When dry a small drop of Canada balsam is to be applied, and the slide put away out of reach of dust to dry for twelve hours. Now place on a hot plate, and apply gentle heat from a spirit-lamp for about ten minutes. Allow it to cool. Take the cover up with a pair of forceps, and bring its balsamed surface in contact with the centre of a warmed slide. The balsam should then run to a neat bevelled edge all round the cover ; should it not do so, warm the slide a little more until it does.

Unselected Polycystina.—Take the forms from the bottle with a glass tube, and spread them on a slide ; dry them over a spirit-lamp. Now clean a cover-glass, fasten it to a slip with your breath, and place a drop or two of balsam on it ; take up some of the polycystina on the point of a knife and place them in the balsam ; stir them well up with a needle and put away for twelve hours. Bake over a spirit-lamp for ten minutes, and while warm stir up again gently with a needle, and spread the forms evenly over the cover. Warm a glass slide, and proceed as directed for unselected diatoms.

Unselected Polycystina as Opaque Objects.—Dry some polycystina on a slide, then take a platinum capsule, put the dried material into it, and heat over a spirit-lamp to a dull red. Clean a cover-glass, fasten it to a slide with your breath, and apply a few drops of balsam. Take up some of the dried forms, put them into the balsam, and stir up with a needle until they are evenly spread over the cover ; put away out of reach of dust for twelve hours, so that the air may escape from the forms. Now place on a hot plate and apply gentle heat for ten to fifteen minutes to bake the balsam. Clean another cover-glass, add a drop or two of balsam to the hardened balsam, and apply the second cover-glass ; warm again, and with a needle press gently on the upper cover until it lies perfectly flat ; then allow to cool, apply a coat of black shellac cement all over one side of the upper cover, and put away to dry. In the meantime take a slip, put it in a turn-table, and run on a disc of black varnish of the same size as the cover ; let this dry, then add a drop of strong gum or glue ; take up the covers with a pair of forceps, and put the blackened side into the glue ; press down with a needle until the glue spreads evenly under the cover, and put away to dry. When dry, finish off with a coat of black cement.

Selected Diatoms and Polycystina.—Take an ounce of distilled water, add 6 or 8 drops of ordinary gum water, and filter. Clean a cover-glass, and place a drop of the diluted gum upon it ; put away to dry.

Spread the diatoms or polycystina on a slip, and dry them over a spirit-lamp. Select the desired forms with a fine brush or bristle, and breathe upon the gummed surface of the cover, and place the forms upon it. When dry, apply a drop of balsam, and put away out of reach of dust for twelve hours. Bake and finish as directed for unselected slides.

In mounting selected polycystina, they must be between two covers; if on a single cover, the forms would be upside down when the cover was reversed. If a transparent mount is desired, the two covers can be fastened to the slide with a drop of balsam. If opaque, the forms must be burnt, and one side must be blackened; in other respects proceed exactly as you would for unselected opaque mounts.

Polycystina may also be mounted in a dry opaque cell. Take a slide, run on a disc of black varnish, and when this has dried fasten a disc of black gummed paper over it. Then take a shallow cell, apply a coat of cement to one side of it, and let it nearly dry; then fasten to the paper disc, and put away to dry. Apply a little dilute gum water to the bottom of the cell, select the specimens, and put them into the gum; if they do not adhere, breathe on the surface of the gum. When all are arranged, put the slide away until everything is quite dry; then add a coat of cement to the upper side of the cell, let it nearly dry, and then apply the cover-glass.

Foraminifera—Unselected Transparent Mounts.—Dry the forms on a slide with the aid of gentle heat, and scrape them off into a bottle of turpentine, in which they must soak until all trace of air has disappeared. Then clean a cover-glass, fasten it to a slide with condensed breath, and apply a few drops of balsam. Pour off the turpentine from the foraminifera, take up some of the forms on the point of a penknife, and put them into the balsam on the cover; stir up with a needle until spread evenly, then put away for twelve hours. Bake gently for ten minutes on a hot plate, cool, apply a drop of fluid balsam, warm a slide over a spirit-lamp, take the cover up in a pair of forceps, and bring the drop of fluid balsam in contact with the centre of the slide, ease down carefully, and press on the upper surface of the cover with the needle-point until it lies quite flat; or if the forms are very delicate, warm the slide again gently until the cover settles down by its own weight. Allow the slide to cool, then clean away exuded balsam with methylated spirit, and apply a coat of cement.

Foraminifera—Opaque Mounts.—Proceed in exactly the same way as directed for mounting dry opaque polycystina ; but if the specimens are unselected, gum the bottom of the cell, dry the forms on a slide, and spread a quantity of them all over the surface of the cell.   Let the gum dry, then shake out all that have not adhered, apply a coat of cement to the upper side of the cell, and when this has nearly dried, apply a cover-glass.

Spicules of Gorgonia or Sea-Fan.—Boil in liq. potassæ until all the material has broken up, then wash away the potash with repeated changes of water, allowing the spicules to settle to the bottom of the tube between each washing.   When cleaned, preserve in a bottle of dilute spirit.   Proceed with the mounting in exactly the same way as directed for transparent unselected polycystina.

Spicules of Alcionium.—Proceed as above.

Sponges.

1. To Show Cell Structure, Flagellated Cell, etc.—Fresh specimens of the calcareous forms—Sycon, for example—should be fixed with osmic acid 1 per cent. solution, washed in distilled water, and placed in absolute alcohol for twelve hours ; then soak in absolute alcohol and ether for a few hours, infiltrate, and embed in celloidin.   Cut sections in a microtome.   Place sections in absolute alcohol for about three minutes, clear in oil of origanum, and mount in Canada balsam.

If preferred, sections of sponges may be mounted in glycerine jelly, but they must be soaked in water for a little while before they can go into the jelly.

2. The Skeleton—(a) *Horny Sponges.*—Boil in liq. potassæ, then wash the spicules well in water, and mount in glycerine jelly or Canada balsam.

(b) *Calcareous.*—Dehydrate small forms in alcohol, clear in clove oil, and mount in Canada balsam in a cell ; or separate the spicules by boiling in liq. potassæ, wash in water, and mount in Canada balsam or glycerine jelly.

(c) *Siliceous.*—Boil in nitric acid, wash well in water, dehydrate, clear, and mount in Canada balsam.

For the types in which siliceous spicules are embedded in horny material, boil in liq. potassæ for a few minutes to disintegrate the tissues, then in nitric acid to clean the spicules, wash well in water, and mount in Canada balsam.

Sections of Sponges.—Harden in methylated spirit, and transfer to equal parts of ether and absolute alcohol for several hours. Then place in a thin solution of celloidin for a day or two, transfer to a thicker solution of celloidin, and soak again for a few hours. Remove from the celloidin on the point of a needle, and hold exposed to the air for a few minutes to allow the celloidin to set around the specimen; then push it off the needle into a bottle of methylated spirit, and soak for a few hours to complete the hardening. Embed in carrot, place in a well microtome, and make the sections. Dehydrate in methylated spirit, clear in oil of bergamot, and mount in Canada balsam.

Sometimes sponge sections are rendered too transparent by mounting in balsam. In such cases, mount in glycerine jelly, but be careful to wash away all traces of alcohol before they go into the jelly.

CHAPTER XIX

DRY MOUNTS

Opaque Cells.—Place a slide in a turn-table, and run a disc of black varnish on its centre; allow this to dry. Take a piece of black paper and punch out a disc of the same size as the one on the slide, and gum it on to the varnish spot. Take a cell, either metal or vulcanite, of the required depth and fasten it to the paper disc with gold size or black shellac cement, and put the slide away until quite dry. Now place a very small quantity of gum on the centre of the paper disc, and put the specimen into it; but take care that the gum does not extend beyond the object, or the appearance of the mount will be spoiled. When the gum has dried, put the slide into the turn-table again, and run a ring of any good cement on the upper surface of the cell, and when this has become about half dry apply a cover-glass, which must be pressed down with a needle-point until it adheres firmly to the cement all round the cell. Put the slide aside for an hour or two, and then run on a good coat of black shellac cement.

Feathers of humming-birds, eggs of butterflies and moths, small microscopic seeds, gills of many fishes, skins of fishes, skins of snakes, and transverse or longitudinal sections of stems of plants, are all mounted as opaque objects in the same manner as above. The former should be arranged in the cell in a group. The gills, skins, etc., should be well washed with distilled water and dried under pressure between two glass slips tied together with twine.

Transparent Cells.—Take a cell of the desired depth and apply a coat of cement to one side of it, and allow it to become very nearly dry. Take a slide and warm it gently over a spirit-lamp ; take up the cell with a pair of forceps and place it on the centre of the slide, the warmth of which should cause the cement of the cell to melt ; if not, warm a little more, and press the cell down gently with a needle-point until it adheres firmly to the slide all round. If the specimen is small it must be fastened in the cell with some gum, as for opaque mounts, then put it away until the gum has dried, apply a cover, and finish off as directed for opaque mounts. Leaves of plants and wings of butterflies should be mounted on a thin slide, so that both sides may be examined. No gum will be required for these specimens, but a piece of the leaf or wing should be cut or punched out as nearly the size of the cell as possible, and a thin cell should be used, so that the cover may rest on the object and keep it flat. In all dry mounts great care must be taken that all the cements used to fasten the objects in position are quite dry before the cover is put on ; if not, any moisture remaining will condense on the under surface of the cover and spoil the preparation.

Opaque Mounts of Pollens.—Make an opaque cell, and apply a thin layer of gum water all over its floor ; then take some perfectly dry pollen and put it in the cell, shake the slide so that the pollen spreads evenly all over the cell, and let it dry. Then apply some enamel to the upper surface of the ring of the cell, and when this is about half dry apply the cover-glass.

CHAPTER XX

FINISHING OFF SLIDES

Canada Balsam—*Quick Method.*—Take a small saucer of chloroform and a soft brush, and carefully wash away the exuded balsam. Allow the slide to dry, then place it in a turn-table and apply a coat of black shellac cement. Let this dry, then wash the slide quite clean with turpentine and apply another coat of cement.

Canada Balsam—*Exposure Method.*—Put the slide into a saucer of methylated spirit, and with a small piece of soft rag gently rub away the excess of balsam; dry the slide with a clean cloth, and apply a coat of any good cement.

Glycerine Jelly.—Put the slide in a saucer of cold water and allow it to soak for a few minutes, then take a penknife and carefully scrape away the jelly from the edge of the cover. Give the slide a good wash in water, and place it in some methylated spirit, which will remove the water. Dry with a clean soft cloth, and apply a coat of black shellac enamel, and when this has dried add another.

Farrant's Medium.—Allow the slide to dry for a few days, then put it into a saucer of water and wash away the excess of medium with a soft brush. Drain off as much water as possible, and, if the cover is firm enough, dry the slide carefully with a soft cloth ; if not, allow all the moisture to evaporate by exposure to the air. When quite dry, put it in a turn-table and apply a coat of cement, and when this has dried add another.

Dry Mounts do not require any washing, but they should have one or two coats of any good cement.

Asphalte and white zinc cement may be used when desired for

balsam or dry mounts, but they are both useless for any of the aqueous or fluid media.

A really good black enamel may be made in the following way :

Dissolve best black sealing-wax in methylated spirit until the solution is as thick as treacle, then mix this with an equal quantity of marine glue ; then if too thick, dilute with a little methylated spirit. This cement has been alluded to in these chapters as shellac cement, and it is the best I know of for general purposes. The black enamel should be kept in a wide-mouthed, stoppered bottle. Should the stopper become fixed, just warm the neck of the bottle over a spirit-lamp; it can then be easily removed.

When a ring is being applied to a slide, the turn-table should not be run too fast, and the extreme point of the brush should only just touch the glass. A thin coat must be run on at first, then give it about ten minutes to dry. A sufficient quantity of cement may then be added to finish the mount, but if too much is applied at first it will overflow.

The most suitable brush for ringing slides is a sable 'rigger' No. 2 in a metal holder ; it should be well washed in methylated spirit after use.

Cleaning off Failures.—During a course of microscopical work many slides will be not worth keeping, but the slips and covers are quite good, and they can be used again. When a batch of failures has accumulated, make a strong solution of Hudson's soap-powder in warm water, and place some of it in two jars. Warm the slide over a spirit-lamp, and with a needle-point push off the cover into one of the jars and put the slip into the other; let them soak for an hour or two, then wash away the soap solution with repeated changes of warm water, and finally pour away all the water and add methylated spirit ; soak for a little while, and then dry with a soft clean rag.

Sometimes slips and covers have a dull, cloudy appearance, which defies all attempts to remove it. When this is the case, make up a solution of hydrochloric acid in methylated spirit (about one part of acid in six of spirit), and immerse the glasses for a few minutes. Wash away the acid with methylated spirit, and dry with a soft rag.

PART III

CHAPTER XXI

AN INTRODUCTION TO THE USE OF THE PETROLOGICAL MICROSCOPE

By FREDERIC J. CHESHIRE, F.R.M.S.

THE petrological microscope is nothing more than an ordinary microscope fitted with certain optical and mechanical adjuncts, by the proper employment of which a more or less complete quantitative and qualitative determination can be made of the optical properties of transparent crystals, as exhibited in polarized light, and thus the identification of such crystals either effected or facilitated. An elementary knowledge at least of the explanation of polarization phenomena is, therefore, imperative for the intelligent and efficient use of the microscope in question —a fact which is too often overlooked by the microscopist.

Polarization of Light.—It is a very curious fact that the understanding of polarization phenomena, when associated with light-waves, should present so many difficulties, when it is remembered that polarized waves are the only forms of wave-motion with which most people are made familiar by their everyday experiences. A pebble dropped into a pond produces a number of small waves, which, starting at the point where the pebble fell into the water, spread outwards in all directions in the form of ever-increasing circles, until they have passed over the entire surface of the pond. But, although these waves travel over the surface, we know that the particles of water concerned in their production do not travel, but simply move up and down. A floating cork, for example, is not carried forward by these waves, but simply rises and falls as each wave passes beneath it. Here, then, we have a case in which waves are propagated from one place to another by the vibratory motion of a large number of particles in one plane only. When waves are transmitted in this way by the motions of particles in a single and constant plane, the wave is said to be polarized. But a simpler and even more instructive example might have been given.

Let us suppose that a long rope is stretched rather loosely between two boys, one of whom, by a rhythmic movement of his hand in an up-and-down direction, produces waves in the rope which pass continuously to the other boy. In this case, too,

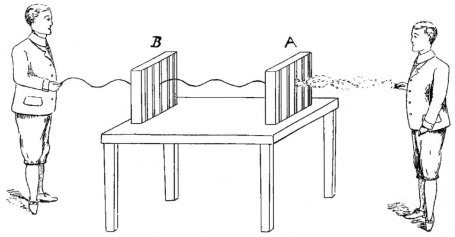

Fig. 66.—Rope Polariscope: Nicols Parallel.

since the vibrations of the particles of which the rope is made up take place in one constant plane only, the resulting waves are polarized ones. But now, as in Fig. 66, let the rope be passed through two gratings, A and B—such, for example, as those used for covering street gullies—in both of which the bars are at first arranged vertically. Now let the first boy start waves along the rope—not by confining the motion of his hand to one plane only, but by changing the direction both rapidly and arbitrarily, so that a jumble of waves is sent along the rope in which the vibrations of the particles take place more or less in all directions. It is clear that these waves are no longer polarized, since the various particles concerned are moving in different planes. This state of affairs would only exist, however, between the first boy and the grating A, for since the bars of the latter are vertical, it is clear that each wave as it falls upon the grating would, in general, be partly stopped and partly transmitted; and, since in the motion transmitted the vibrations would take place in one constant plane only, it would be polarized. The grating A would thus act as a polarizer, and we should have a succession of polarized waves passing from it to the second grating B, and thence, since the bars of B are also arranged vertically, to the second boy. Had the bars of the grating B been arranged horizontally—that is, at right angles to those of the first—it is clear that the polarized waves falling upon the second grating would be stopped

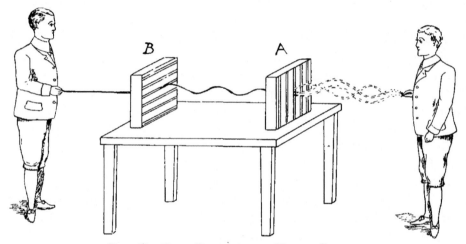

FIG. 67.—ROPE POLARISCOPE: NICOLS CROSSED.

as in Fig. 67, but in all intermediate positions of these bars, between the vertical and horizontal, the waves from A would be partly stopped and partly transmitted with the plane of vibration changed. Finally, were it required to determine the plane in which the vibrations were taking place in the waves falling upon B, it would only be necessary to rotate this grating into such a position that no waves were transmitted. The bars would then be at right angles to the sought-for direction. In this way the grating B could be used to test the polarized condition of the waves falling upon it—*i.e.*, as an analyzer.

Polarized Light.—Ordinary light may be looked upon as consisting of waves transmitted through the ether of space by the to-and-fro motions of the particles of that ether in all directions across the line of march of the waves ; but when such a beam of light falls upon a nicol prism, it is polarized—that is, after passing through the prism, the light-waves are found to have their vibrations in a single and constant plane. The nicol prism, therefore, under these circumstances acts upon light-waves in an analogous way to that in which the grating A acted upon the rope-waves. Similarly, the polarized light produced by the action of one nicol prism being allowed to fall upon a second one similarly arranged, will pass through it, just as the waves from the grating A passed through the grating B, when similarly arranged, as in Fig. 66. The second nicol also being turned through a right angle from the position just described, completely stops the light falling upon it from the first one, just as the second grating B stopped the waves from A in the position shown by Fig. 67. Finally, in all intermediate positions of the second nicol the light will be more or less completely transmitted, but with its plane of vibration changed. To sum up, therefore, we may say that—

1. In ordinary or common light, such as sunlight, the vibrations of the ether particles concerned in the transmission of the light-waves take place in every possible direction across the direction in which the light is moving.

2. A nicol prism (or its equivalent) is a kind of optical grating which reduces the vibrations of ordinary incident light to a single direction or plane only, and is thus said to act as a polarizer.*

3. A nicol prism, when polarized light falls upon it, transmits it in general more or less completely with its plane of vibration changed, but in the particular case when it is so arranged that the direction in which it allows ether vibrations to take place is at right angles to the direction in which the vibrations in the incident polarized light are taking place, it stops the incident light altogether, and thus acts as an analyzer.

Double Refraction.—The peculiar optical property possessed by most crystals in virtue of which their examination and differentiation in a petrological microscope becomes possible is

---

* This action of a nicol prism must not be confounded with that of the grating employed in spectrum analysis, which acts, of course, in an entirely different way.

known as double refraction, and is due to what might very appropriately be called 'optical grain.' A piece of wood has a grain which is usually, it is true, apparent to the eye; but even in the absence of such evidence the fact could soon be determined experimentally, as by attempting to split the wood with a hatchet in various directions. Rock crystal, the clear and transparent variety of quartz, sometimes takes the form of long prisms of hexagonal cross-section, as shown by Fig. 68. Now, it is found that if a slice be cut lengthwise from such a crystal, as indicated at AA, and smeared with wax on one of its faces, the application of heat to a point on that face will cause the wax to melt. The area over which the melting occurs will not, however—as in the case of glass, for example—take a circular form, but an elliptical one, with the major axis of the ellipse parallel to the geometrical axis of the crystal. This experiment shows that heat is transmitted more rapidly along the crystal than across it. In a slice cut across the axis, as at BB, the melted area takes a circular form, showing that across the

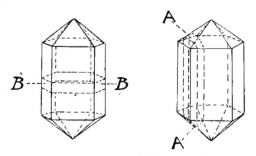

FIG. 68.—CRYSTALS OF QUARTZ.

crystal heat is transmitted equally in all directions. Rock crystal is thus shown to be possessed of a kind of grain running in the direction of its length. In consequence of this grain, which is optical as well as thermal, it is found that if a beam of light be allowed to fall normally upon the face of such a slice of rock crystal, the latter will be found to act as a kind of double grating, sifting and transmitting the incident light as two beams, in one of which the vibrations take place only in the direction of the length of the crystal, and in the other beam across it only. A nicol prism, it may be remarked, is a crystal of Iceland spar in which one of the two beams just referred to is thrown out of the way by reflection on an artificial interface. A beam of light passed along the axis of a quartz crystal is not split up into two beams, but in any other direction it is; hence rock crystal is what is known as a uniaxial crystal. In other crystals, as mica and selenite, there are two directions in which light passes without being split up; these are therefore known as biaxial crystals.

Now the two polarized beams into which light is, in general, split up on its passage through a crystal, travel with different speeds, and in most cases in slightly different directions—hence the term 'double refracting.' In the case of rock crystal, for example, the beam in which the vibrations occur along the axis of the crystal travels more slowly than the beam in which the vibrations occur across it.* Let such a slice of rock crystal be placed between crossed nicols, and arranged with its axis inclined to the vibration-directions P and A of the polarizer

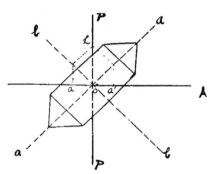

FIG. 69.—SECTION OF QUARTZ CUT PARALLEL TO AXIS.

and analyzer respectively, as shown by Fig. 69. Then polarized light coming upwards from the polarizer towards the observer will be resolved by the crystal into two beams with vibrations in rectangular planes and passing with different speeds. These two beams, falling upon the analyzer, will each again be resolved into two beams—one with horizontal and the other with vertical vibrations. The first of these will pass the analyzer, the second will be stopped. But further than this, the light which passes the analyzer and emerges as a single polarized beam is compounded of two beams of equal intensities, one of which passed through the crystal with vibrations along aa, and the other with vibrations along bb; and, since these beams travelled at different speeds, it follows that upon being compounded by an analyzer the waves in one will be more or less out of step with those in the other, with the result that interference will take place, and the waves corresponding to any colour trans-

---

* This statement is made upon the usual assumption that the vibrations of the ether take place in a direction at right angles to the plane of polarization. It should also be remembered that light polarized by reflection at a plane glass surface is *defined* as being polarized in the plane of reflection. It follows therefore from the given assumption and the definition that the vibrations in the polarized reflected light are executed in directions parallel to the surface of the glass.

mitted in opposite phase by the two paths will be destroyed, leaving the transmitted beam coloured. Thus, when the crystal is of such a thickness that the waves by the slow path emerge 550 micromillimetres (550 $\mu\mu$) behind those passing by the fast path, green light with a wave-length equal to this quantity will be cut out, leaving the transmitted beam of the complementary colour red. As shown by Fig. 69, an amplitude *oh* in the polarizer is, after being resolved along each of the two vibration-directions in the crystal and the vibration-direction (horizontal) of the analyzer, represented by two equal amplitudes *oa* and *oa′*. Different colours produced in this and analogous ways have been very carefully studied, and are set out in the following table. This succession of colours would be produced by a wedge of selenite with its vibration-directions adjusted diagonally as in Fig. 69, between crossed nicols, the thickness of the wedge increasing from nothing up to about 0·2 mm. The first column

NEWTON'S COLOUR SCALE ACCORDING TO QUINCKE.

Retardation in Micromillimetres.	Interference Colour between Crossed Nicols.	Order.	Retardation in Micromillimetres.	Interference Colour between Crossed Nicols.	Order.
0	Black.	First.	843	Yellowish-green.	Second.
40	Iron-grey.		866	Greenish-yellow.	
97	Lavender-grey.		910	Pure yellow.	
158	Greyish-blue.		948	Orange.	
218	Clearer grey.		998	Bright orange-red.	
234	Greenish-white.		1,101	Dark violet-red.	
259	Almost pure white.				
267	Yellowish-white.		1,128	Light bluish-violet.	Third.
275	Pale straw-yellow.		1,151	Indigo.	
281	Straw-yellow.		1,258	Greenish-blue.	
306	Light yellow.		1,334	Sea-green.	
332	Bright yellow.		1,376	Brilliant green.	
430	Brownish-yellow.		1,426	Greenish-yellow.	
505	Reddish-orange.		1,495	Flesh colour.	
536	Red.		1,534	Carmine-red.	
551	Deep red.		1,621	Dull purple.	
565	Purple.	Second.	1,652	Violet-grey.	Fourth.
575	Violet.		1,682	Greyish-blue.	
589	Indigo.		1,711	Dull sea-green.	
664	Blue (sky-blue).		1,744	Bluish-green.	
728	Greenish-blue.		1,811	Light green.	
747	Green.		1,927	Light greenish-grey.	
826	Lighter green.		2,007	Whitish-grey.	

gives the retardation—*i.e.*, the distance in micromillimetres which one beam emerges behind the other after passing through the crystal—whilst the second column gives the colour corresponding to such retardation.

Rotary Polarization.—Certain crystals possess in a certain direction the remarkable power of rotating or twisting the plane of vibration of a polarized beam passing through them. Thus, if a slice of rock crystal 1 mm. thick, cut from the crystal normal to the axis as at BB, Fig. 68, be placed between crossed nicols in white light, it is found that the analyzer no longer stops the light, and that no position can be found for it in which it does stop it. In the case of sodium light, however, it is found that a rotation of the analyzer from its crossed position with respect to the polarizer through an angle of 22° again establishes darkness. In some instances this necessary rotation has to be made in the direction of the hands of a clock, from the observer's point of view, about the direction in which the light is passing to the eye, whilst in others it has to be made in the opposite direction. In the first

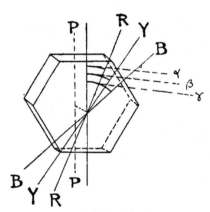

FIG. 70. — SECTION OF QUARTZ, NORMAL TO AXIS, TO SHOW ROTARY POLARIZATION.

case the rotary polarization is said to be right-handed, and in the second case left-handed. If, then, a beam of polarized sodium light, in which the vibrations are vertical, is allowed to pass along the axis of a quartz crystal, the plane of vibration will not remain vertical, but will be gradually rotated or twisted about that axis at the rate of 22° per mm. of length of the crystal. For different colours in the incident polarized white light, the rate of turning is different. Red light, for example, has its plane of vibration twisted at the rate of 13° per mm., whilst blue, on the other hand, is twisted through as many as 33° in the same length. Thus, if Fig. 70 be taken to represent this action in a slice of right-handed quartz of the specified thickness in incident white light, polarized with its vibrations along the line PP, and passing upwards to the eye, the vibration planes for the red, yellow, and blue will be twisted into the directions RR, YY, and BB respectively, through angles a, β, and γ, equal to 13°, 22°, and 33° respectively. The analyzer, therefore, set originally parallel with the polarizer and rotated in the direction of the hands of a clock, would allow in succession the colours red, yellow, and blue to pass in predominance to the eye, so that the crystal would appear to change in colour during the rotation of the analyzer.

Optical Adjuncts for the Petrological Microscope.

Crystallographic determinations are very much facilitated by the employment of a number of optical adjuncts, the more indispensable of which are set out below.

Mica Quarter-Wave Plate.—This consists of a cleavage plate of mica of such a thickness that sodium light, with a wave-length of 589 $\mu\mu$, in passing through it, along one of the two vibration-directions possible in a double-refracting crystal, emerges a quarter of a wave-length behind that passing through by the other rectangular vibration-direction. These vibration-directions are often, therefore, referred to as the 'fast' and 'slow' directions to differentiate them. If a quarter-wave mica then be placed on a crystal section, so that similar directions in the two sections are parallel to one another, the effect is the same as that which would have been obtained by increasing the thickness of the crystal section, and if the latter should be between crossed nicols with its vibration-direction inclined at 45° to the vibration-directions of the nicols, its colour would rise in Newton's scale—*i.e.*, correspond to a greater retardation. By superposing the mica with its fast direction parallel to the slow direction of the crystal section, the effect would be the same as that which would have been obtained by decreasing the thickness of the crystal section, so that in this case the colour would descend in Newton's scale—*i.e.*, correspond to a less retardation. By the use of a quarter-wave mica in this way the fast and slow directions of crystal sections are differentiated.

Selenite or Gypsum Plate.—When it is required to differentiate the fast and slow directions of a crystal section with small bi-refracting power, a selenite plate of such a thickness as to give a rose colour between crossed nicols is employed. Superposed upon a crystal section with similar directions parallel, the colour changes to blue, whilst the placing of fast on slow changes the colour to red.

Klein Quartz Plate.—A plate of quartz 3·75 mm. thick, and cut at right angles to the axis of the crystal, gives between crossed nicols, and in virtue of its rotary polarizing power, a purple colour. In this case the orientation of the plate does not affect either the intensity or the colour of the light transmitted.

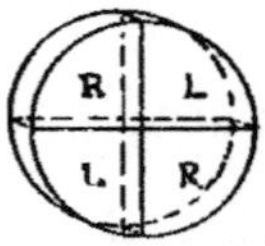

Bertrand Plate.—This plate is made up of four quadrantal sectors of alternately right- and left-hand quartz, cut at right angles to the axis of the crystal, and 2·5 mm. thick.

FIG. 71.
BERTRAND PLATE.

Fedorow Mica-Steps.—This is built up by superposing some sixteen strips of quarter-wave mica, all with similar directions parallel, in such a way that each strip is about 2 mm. shorter than the one immediately below it.

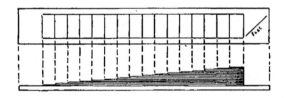

FIG. 72.—MICA-STEPS.

Sixteen steps are thus formed, which effect in succession retardations, increasing by a quarter-wave at each step, commencing with a quarter-wave, and finishing with four waves.

Quartz Wedge.—A thin slice of quartz cut parallel to the crystallographic axis is ground into the form of a thin wedge. Could such a wedge be ground to an infinitely thin edge, it would give at this edge, when oriented with its vibration-directions in diagonal adjustment between crossed nicols, the black of Newton's colour scale, followed by all the colours of the scale in ascending order in passing along the wedge to the thick end. Wedges giving the first six orders of Newton's scale, or some smaller number if required, are thus made. To avoid the necessity for grinding a very thin edge, the quartz plate A from which the wedge is to be made, is cemented to a thin plate of selenite B, with the fast direction of one parallel to the slow direction of the other.   By this device it is made easy to get the starting black at the thin end of the quartz without reducing that end to a less thickness than the selenite foundation-plate possesses.   Sometimes this foundation-plate is made to give the sensitive rose tint of the first or second order, and is made to project for a short distance beyond the thin end of the quartz. The wedge should carry a scale along its length, from which the retardation in micromillimetres at any point of the wedge, and for any given tint, can be determined.

It will be found that the optical adjuncts referred to above, as produced by different makers, are unfortunately not uniformly

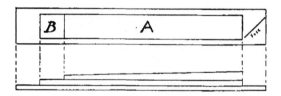

FIG. 73.—QUARTZ WEDGE.

mounted as regards the orientation of their vibration-directions with respect to the length of the plate or wedge.   Sometimes the fast direction is coincident with the length of the plate, sometimes it is across it, and sometimes it will be found inclined at 45° to the length.   The last disposition has the advantage that by inverting the plate in the cross slot in the tube of the microscope, the optical superposition of fast on fast can be changed for slow on fast without any difficulty.

The Construction of the Petrological Microscope.—Fig. 74 shows a first-class modern instrument.   The sub-stage polarizer is associated with a triple condensing system for convergent light, two lenses of which can be turned to one side when plane-polarized light is required.   The stage is rotatable, and graduated to read to a tenth of a degree with the help of a vernier.   The collar to which the objective is secured is fitted with centring screws, and is made with a slot into which the usual mica and gypsum compensators can be introduced.   The

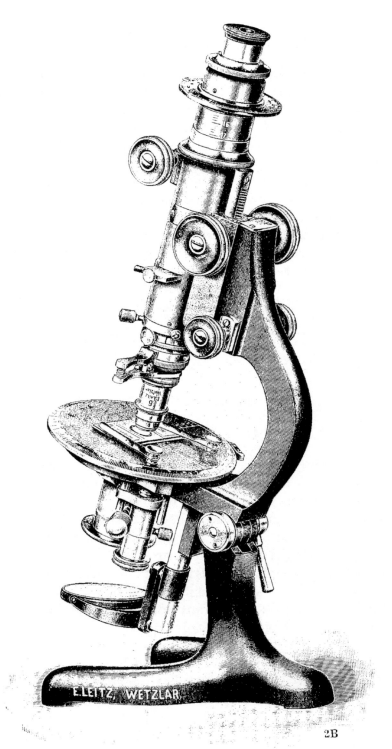

2B

Fig. 74.—A Modern Petrological Microscope.

analyzer fitted above the objective can be pushed radially into and out of action, and it can further be rotated, when required, through an angle of 90° or less about the axis of the microscope body-tube, and clamped in position.   The Bertrand lens slides into position near the middle of the length of the tube.   The draw-tube is operated by a rack and pinion, and an auxiliary analyzer with divided circle and a slot for compensators may be fitted over the eyepiece.   The fine adjustment head is graduated to read directly to a thousandth of a millimetre.

Preliminary Adjustment of a Petrological Microscope.— Before any work is done the following adjustments should be carefully tested, and, if necessary, made :

1. Centring of the objective.

2. Rectangularity of the cross-wires in the eyepiece.

3. Rectangularity of the vibration planes of the two nicol prisms.

4. Parallelism of the cross-wires to the vibration planes of the two nicol prisms when the latter are crossed.

Centring of the Objective.—This is done in microscopes of the usual type by the manipulation of two radial set-screws acting against the collar in which the objective is secured.   A slide should be placed upon the stage, and a prominent point or feature of it adjusted to the intersection of the cross-wires. Upon a complete rotation of the stage the point selected will describe a small circle in the field of view ; consequently, when the stage has been rotated through 180° only, the point selected will have its maximum displacement from the intersection of the cross-wires.   Stop the rotation, therefore, at this point, and turn the adjusting screws so as to move apparently the inter- section of the wires half-way towards the selected point.   The adjustment will now be found to be very nearly correct.   Repeat the operation until it is quite so.

Rectangularity of the Cross-Wires in the Eyepiece.— Place a slide with a fine straight line ruled upon it on the stage, and adjust it until the projected image of the line coincides with one of the cross-wires.   Note the angular position of the stage, and rotate it carefully through a right angle.   The projected image of the line should now be parallel to the second cross- wire.   If the centring of the objective has been first effected as above, the projected image in the second case will coincide with the second cross-wire.

Rectangularity of the Vibration Planes of the Nicol Prisms.—Darkness of the field is not a sufficiently delicate test for this adjustment, but, if the necessary adjunct—a Bertrand quarter-quartz plate—is not available for a more delicate adjustment, darkness should be obtained a number of times by rotation of the polarizer alternately in opposite directions. The mean of the various readings should be taken as the true one. To obtain a better result, place a Bertrand plate upon the stage, set the polarizer to zero, and slide the analyzer into position. If the vibration planes of the two nicols are accurately at right angles to one another, the quadrants of the Bertrand plate will appear to have the same tint. Otherwise, the colours of adjacent quadrants will not match, in which event the polarizer must be rotated until they do match, when the necessary zero correction should be read off on the polarizer. If the polarizer is not fully graduated so as to allow of this being done, a fine vertical line should be drawn across the junction of the polarizer mount and the sleeve into which it is pushed. Better still, if possible, the polarizing prism should be rotated in its mount, until the latter being at zero, the adjustment is correct.

Parallelism of the Cross-Wires to the Vibration Planes of the Crossed Nicols.—A needle-shaped crystal such as anhydrite (anhydrous sulphate of calcium, crystallizing in the orthorhombic system), in which one of the directions of extinction is parallel to the long edges of the crystal, should be placed on the stage between crossed nicols, and rotated until extinction is obtained. Pull the analyzer out, when the crystal should be seen ranged parallel to one of the cross-wires. Turn the crystal over on the stage and repeat.

Examination and Identification of the Crystalline Constituents of Rock Sections.

As this chapter does not profess to be anything more than an introduction to the use of the petrological microscope, no attempt will be made to describe the complete and systematic examination usually made of crystal sections by expert mineralogists.   Some of the simpler determinations only will be indicated.   Suppose that an angular crystal which lights up and darkens between crossed nicols upon rotation of the stage is to be examined, we could proceed to determine (1) the angles between the sides of the crystal; (2) the angular positions of the extinction-directions with respect to the sides; (3) the differentiation of these extinction-directions into fast and slow; and (4) the retardation of the section.

To Measure the Plane Angles of a Crystal Section.— Neither nicol is necessary.   Adjust the section on the stage until one of the sides of the section is projected along a cross-wire.   Take the stage (angular) reading.   Rotate the stage until the second side is brought into alignment with the same cross-wire.   Take a second stage reading.   The difference between these two readings gives the desired angle.

To Determine the Angular Positions of the Extinction-Directions.—Cross the nicols and adjust the section until a side of the crystal coincides with a cross-wire.   Take the stage reading.   Rotate the stage to extinction.   Take the stage reading again.   The difference is the desired angle.   Repeat, and take the mean value of the results.   To make a more accurate determination, advantage is taken of the fact that the four sectors of a Bertrand quarter-quartz plate, placed in the eyepiece between accurately crossed nicols, will appear of one uniform tint whenever a bi-refracting plate on the stage is rotated into such a position that its vibration-directions are parallel to those of the crossed nicols.   This method necessitates the employment of an analyzer above the eyepiece.

To Differentiate the Extinction-Directions.—This may be done by the use of a mica quarter-wave plate in the way already described.   When, however, the bi-refracting power of the section being examined is very small, it is better to employ a sensitive selenite plate (so-called red of the first order).   When this is introduced into the cross-slot just over the objective, so that its fast direction is parallel to the fast direction of the crystal section, the rose colour changes to a blue; whilst when the fast direction is superposed on the slow of the section, the colour becomes a bright red.

Retardation.—This quantity is most simply determined by the use of the quartz wedge or the mica-step compensator. Unfortunately, however, these cannot be used satisfactorily in the usual slot over the objective, because in neither case, in the final adjustment, is one thickness only of the compensator operative. In the mica-step, for example, two or more steps must be interposed in the path of the light rays proceeding from the objective to form the image in the eyepiece, whilst in the case of the quartz wedge quite an appreciable fraction of the total length must be interposed. Further, in the latter case, the retardation scale cannot be used since it is not in focus. When a very low-power objective is sufficient, the compensators can be superposed on the section on the stage, and the retardation determined directly by the position, in the case of the

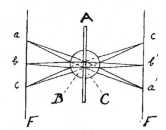

FIG. 75.—THE ACTION OF A CONVERGENT SYSTEM.

quartz wedge, of the black band on the retardation scale when the fast direction of the wedge is superposed on the slow direction of the section. In the case of the mica-step the retardation may be equal to that of an integral number of steps, but more generally it falls between two of these, and an estimate of its value has to be made. These compensators should always, if possible, be used in the focal plane of the eyepiece. In that event, of course, the usual analyzer must be thrown out of action and one placed instead over the eyepiece.

Examination in Convergent Light.—To understand the optical action which is taking place in the microscope when it is being employed for the examination of sections in convergent light, it will be better to consider first the simple case (Fig. 75), in which a plate of bi-refracting crystal A cut at right angles to the axis—a plate of calcite, say—is interposed between two nearly hemispherical lenses, B and C. Further, let plane-polarized light, no matter how produced, start from the point *b*, on the axis of the system and in the principal focal plane F of the lens B, and passing through the lens B, plate A, and lens C, be brought to a focus by the latter at the point *b'* in the focal plane F' of the lens C. Light from points *a* and *c* will similarly be brought to a focus in points *a'* and *c'* respectively. Now, it will be observed that in each of these three cases the light that actually passes through the plate A is in the form of parallel rays, but that the inclination which any particular bundle of parallel rays makes with the axis of the crystal —the line *bb'*—depends upon the distance *bc* or *ba*. In the plane F', therefore, it follows that the light falling in the circular line struck with a radius *b'c'*, around the point *b'*, is light, the whole of which has passed through the crystal plate at the same angle of inclination to the axis. In the plane F', therefore, we get the familiar interference figure which, when looked at through a crossed nicol, appears as a number of concentric rainbow-tinted rings, with a black cross marking them off into quadrants.

In the actual microscope the condenser B, Fig. 76, functions as the lens B of Fig. 75, and the objective C as the lens C of Fig. 75. The interference image shown as being focussed in the upper focal plane of the objective, is projected by the Bertrand lens D and the field-lens E of the eyepiece, into the stop-plane of the latter, and again by the eye-lens F on to the retina of the eye of the observer. The analyzer is shown fitted between the Bertrand lens and the objective. The eyepiece and the Bertrand lens thus act together as a low-power compound microscope to magnify the figure in the upper focal plane of the objective C. In the absence of the Bertrand lens the eyepiece projects the interference figure

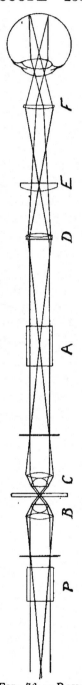

FIG. 76.—RAY-DIAGRAM FOR A MICROSCOPE ARRANGED FOR

satisfactorily observed with a powerful pocket magnifier. In the latter case a small stop may be placed in the stop-plane of the eyepiece to cut off all light except that which has passed through the crystal under examination.

Plate I., from the atlas of the late Dr. Hausewaldt, shows the interference figures in convergent sodium light and between crossed nicols of sections of arragonite—the first pair due to a specimen ½ mm. thick, the second pair due to one of 4 mm. thickness.

Figs. 77 and 79 show the figures when the extinction-directions of the crystal are adjusted parallel to those of the nicol; Figs. 78 and 80 when those directions are adjusted diagonally.

The attention of the reader desirous of further information is directed to a paper by Dr. John Evans in the Proceedings of the Geologists' Association, vol. xxi., part 2, 1909, on ' The Systematic Examination of a Thin Section of a Crystal with an Ordinary Petrological Microscope.' It is to be regretted that this invaluable brochure has not been published in a more accessible form. The following textbooks may also be referred to—viz., ' Traité de Technique Minéralogique et Petrographique,' by Duparc and Pearce, Leipzig, 1909; and ' Anleitung zum Gebrauch des Polarisationsmikroskops,' by Weinschenk, Freiburg im Breisgau, 1906.

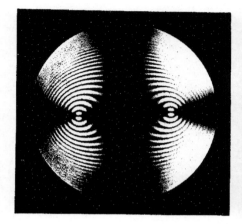

FIG. 77.

FIG. 78.

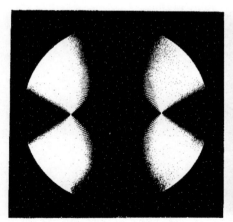

FIG. 79.

FIG. 80.

INTERFERENCE FIGURES OF ARRAGONITE. (HAUSEWALDT.)

[To face p. 236.

CHAPTER XXII

ROTIFERA*

BY C. F. ROUSSELET, F.R.M.S.

ARRANGED in five sections as follows ·
1. Specimens likely to be found during each month of the year ;
2. Collecting-grounds near London ;
3. Methods of collecting, preliminary examination, and keeping ;
4. Apparatus for microscopic examination ;
5. Preserving and mounting.

1. Specimens likely to be found during each Month of the Year.

JANUARY.

January is the most severe month of the year, and lakes and ponds are often frozen over or difficult to approach. Microscopic pond-life, though less abundant than in the spring and autumn, is, nevertheless, nearly always present, even under the ice many inches thick. All the following species of rotifers have been taken in January in and near London; but no doubt a great many more could be found by systematic search · *Asplanchna Brightwellii* and *priodonta ; Anuræa aculeata* and *cochlearis ; Brachionus pala* and *angularis ; Notholca scapha ; Euchlanis deflexa* and *hyalina ; Rotifer macrurus* and *vulgaris ; Polyarthra platyptera ; Synchæta pectinata, tremula,* and *oblonga ; Conochilus unicornis ; Cœlopus porcellus ; Diaschiza lacinulata* and *ventripes ; Proales decipiens* and *petromyzon ; Diglena forci-*

* Originally published in *Knowledge*, and reproduced by permission.

pata ; Dinocharis pocillum ; Monostyla cornuta ; Colurus caudatus; Melicerta ringens ; Limnias ceratophylli ; Œcistes crystallinus ; Floscularia cornuta ; and *Stephanoceros eichhornii.* Diaptomus and Cyclops and their larvæ are abundant, whilst Water-fleas are almost absent. Aquatic vegetation having died down, the fixed forms of rotifers and Infusoria should be looked for on the rootlets of trees growing near the edge of the water. Floscules and Melicerta were once found covering such rootlets very thickly. January seems to be the time when the males of Stephanoceros and other tube-dwellers are found, and their presence is often betrayed by the thick-shelled, fertilized, resting eggs in some of the tubes, and numerous smaller male eggs in others.

FEBRUARY.

In the early part of the year, when the weather is still cold and ponds are covered with ice, some Infusoria may be found in abundance, particularly the various species of Vorticella— *Carchesium polypinum, Zoothamnium arbuscula, Epistylis flavicans* —attached to submerged rootlets.

Rotifera to be looked for in lakes and ponds, particularly duck-ponds : *Anuræa aculeata, Anuræa cochlearis, Asplanchna priodonta* and *Brightwellii, Notholca scapha, Polyarthra platyptera, Euchlanis deflexa, Synchæta tremula.* The water-plants having mostly died down, the following fixed forms are found attached on Anacharis, or on submerged rootlets of plants, or on trees growing near the edge of ponds and lakes : *Melicerta ringens, Limnias ceratophylli, Stephanoceros eichhornii, Floscularia cornuta,* and others ; *Œcistes crystallinus* and others.

MARCH.

The same species as those mentioned for February are still to be found, and in greater abundance. Some new Infusoria will have made their appearance, such as *Stentor polymorphus,* which will be found covering the rootlets of Duckweed and other submerged plants, *Peridinium tabulatum* and the free-swimming colonies of *Synura uvella,* etc. Then the very minute and beautiful colonies of Collared Monads, *Godosiga umbellata,* and other species of this group may be looked for, attached to the stems of Vorticella trees.

All the Rotifera forming the winter fauna will become very abundant in March, and as the food-supply in minute Algæ and Infusoria increases, fresh species make their appearance with every rise of temperature. The following additional species may be looked for: *Brachionus angularis ; Notholca acuminata, spinifera,* and *labis; Euchlanis oropha ; Dinocharis pocillum ; Diaschiza lacinulata ; Proales decipiens* and *petromyzon ; Monostyla cornuta; Diglena forcipata ; Rotifer vulgaris.*

April.

All species of Infusoria and Rotifera mentioned as occurring in March are likely to become more abundant in April, which is one of the best months for collecting. The ponds are full of water, whilst they have become approachable, and Daphnias and Cyclops have not yet crowded out the rotifers, as sometimes occurs later on. *Volvox globator* may be looked for, together with the little parasitic rotifer, *Proales parasitica,* inside the green spheres.

Of larger Infusoria, *Bursaria truncatella, Chœnia teres, Amphileptus gigas,* and *flagellatus* will be found, and, of course, crowds of *Euglena viridis.*

Of Rotifera, *Synchœta pectinata* will be abundant, and *Asplanchna priodonta* and *Brightwellii* will have made their appearance in larger lakes and canals ; also *Brachionus pala, quadratus,* and *Bakeri ; Euchlanis triquetra* and *hyalina; Triarthra longiseta, Diaschiza semiaperta ; Rhinops vitrea, Pterodina patina, Mastigocerca bicornis,* and many others.

May.

All the various pond organisms that die down in winter and in various ways produce protected germs to tide over this, for them, unsuitable season, will now have come to life again and begin to multiply at an increasing rate. Many kinds of Desmids should be found in shallow, mossy pools, or along the edge of rivulets. Among Protophyta and Protozoa the green spheres of

Volvox globator will be found in many localities more or less abundantly, and the various kinds of Acineta should be looked for in quiet, undisturbed waters, where many kinds of free-swimming Infusoria will also be found.

Of Rotifera there are few species which may not be found in May. At one excursion of the Quekett Club to Totteridge in the middle of May forty different species were obtained. To mention only a few : *Notops brachionus*, one of the most attractive rotifers, will have made its appearance ; then various kinds of Anuræa, Asplanchna, Brachionus, Cœlopus, Cathypna, Diaschiza, Euchlanis, Furcularia, Mastigocerca, Metopidia, Pterodina, Synchæta, Scaridium, Stephanops ; also *Stephanoceros eichhornii*, Floscules, Melicerta, and Limnias in abundance.

On rootlets of trees growing near the edge of ponds and lakes will probably be found various kinds of Polyzoa : *Fredericella Sultana*, *Paludicella*, and *Plumatella repens*.

JUNE.

If the months of April and May are abnormally cold, pond organisms which usually make their appearance in May are likely to be retarded, and will only come on in June. There are, however, summer forms which hardly ever occur earlier than June, and the most interesting of these amongst rotifers is *Pedalion mirum*, with its six arthropodous limbs ; *Synchæta stylata*, with its long-spined floating eggs, and *Synchæta grandis*, the largest species of this genus, may also now be looked for in lakes and water reservoirs, as well as the rare free-swimming *Floscularia pelagica*. In the same waters will be found two free-swimming colonies of Vorticella : *Epistylis rotans* and *Zoothamnium limneticum*. In June it often happens that certain water-fleas, Daphnia and Bosmina, also Cyclops and their larvæ, increase to such an extent as to render the existence of free-swimming rotifers almost impossible in these waters, and the latter consequently disappear, though they may have been swarming a few weeks earlier. In ponds, however, where this does not occur, rotifers of many genera may be found, and attached to submerged water-plants *Lacinularia socialis* and *Megalotrocha albo-flavicans* should be looked for, whilst in reedy ponds the free-swimming spheres of *Conochilus volvox* may occur. Mossy pools, in addition to their special rotiferous fauna of Philodina, Callidina, Adineta, Cathypna, Distyla, and Monostyla, will also contain water-bears and shelled Rhizopods, such as Diflugia and Arcella and numerous free-swimming Infusoria. Polyzoa, such as *Plumatella repens*, *Fredericella Sultana*, *Lophopus crystallinus*, and *Cristatella mucedo*, though not common, should be abundant in suitable localities.

JULY.

Collecting in July is usually not so profitable as one would expect, because as a rule most of the shallow ponds are dried up by this time, or have been reduced to a muddy swamp, and in the others Crustaceans, Cladocera, and Cyclops have multiplied to such an extent as to leave little room for the more interesting forms of pond-life.

Pedalion mirum should be looked for in large and small lakes, as it will probably have greatly increased in numbers. The somewhat rare and very large *Asplanchna amphora* and *ebbesbornii*, as well as *Asplanchnopus myrmeleo*, are summer forms which occur at this season. Other rotifers that appear in warm weather are : *Dinops longipes; Triphyllus lacustris; Notops clavulatus; Scaridium eudactilotum* and *longicaudum;* then the free-swimming *Lacinularia natans* and *Conochilus volvox;* also the fixed *Lacinularia socialis* and Megalotrocha, which are found attached to submerged water-plants. All these are very beautiful objects under the microscope, but by no means common.

Volvox globator will certainly be found in abundance now in secluded ponds, and inside the green spheres the little parasitic rotifer, *Proales parasitica*, should be looked for.

The Polyzoa, mentioned last month, will have become more abundant where they occur ; undisturbed ornamental lakes and canals are the best places to find them in.

AUGUST.

For the collector of Cyclops, Diaptomus, Water-fleas and aquatic insect larvæ, August is a very capital month ; not so, however, for the collector of the more interesting Infusoria and Rotifera, which are usually quite crowded out by the more vigorous Crustaceans in the few remaining ponds and pools not wholly dried up. In larger lakes, however, it is possible to find occasionally a number of interesting forms, particularly free-swimming rotifers, such as *Asplanchna priodonta* and *Brightwellii*, *Synchæta pectinata*, and the rarer summer forms, *Synchæta stylata* and *grandis*. Where a 'green' pond can be found full of the flagellate Infusorian *Euglena viridis*, there are usually present also a number of rotifers, such as *Hydatina senta*, *Eosphora aurita*, *Diglena biraphis*, etc., feeding on the Euglena.

In shady forest pools, overgrown with Sphagnum, quite a peculiar fauna of moss-haunting rotifers will be found, particularly various species of Callidina, Distyla, Metopidia, Cathypna, in addition to numerous interesting Rhizopods with shells of various forms. In similar ponds the large but very rare rotifer, *Copeus spicatus*, should be looked for. Of other rotifers that may be met with in lakes, more or less abundantly, the following can be mentioned : *Brachionus pala ; Anuræa aculeata, brevispina*, and *hypelasma ; Dinocharis pocillum ; Euchlanis triquetra, hyalina*, and *oropha ; Mastigocerca bicornis, elongata*, and *stylata ; Polyarthra platyptera ; Synchæta tremula* and *oblonga ; Pedalion mirum*, and many others.

September.

In normal years many of the dried-up ponds begin to fill up again in September, and become then most prolific in infusorian and rotiferous life, because the disturbing Crustaceans, Cyclops, and Cladocera have been to a large extent eliminated. But also in larger ponds and lakes, which do not dry up, the Crustaceans decrease in numbers and give the Rotifera and Infusoria a fresh chance of increase. The following free-swimming forms may often be collected in immense numbers : *Asplanchna priodonta, intermedia*, and *Brightwellii ; Triarthra longiseta ; Polyarthra platyptera ; Synchæta pectinata, tremula*, and *oblonga ; Anuræa aculeata* and *cochlearis ; Brachionus angularis ; Pedalion mirum ; Conochilus unicornis*, and the much rarer *Floscularia pelagica*. Of the fixed forms, *Limnias ceratophylli* and *annulatus*, *Cephalosiphon limnias, Lacinularia socialis, Melicerta ringens* and *conifera* should be looked for on submerged water-plants, such as Anacharis, Ceratophyllum, Nitella, and on the rootlets of Duckweed.

Polyzoa such as Plumatella, **Lophopus**, Cristatella, should be found in abundance in disused canals and backwaters of rivers and the larger lakes, from which they could be dredged with a loaded hook and line.

It may be taken as a general rule that all the more interesting forms of pond-life become more abundant in September, provided only that the weather is not too hot, but tempered by repeated showers to fill the dried-up ponds with a fresh supply of rain-water.

OCTOBER.

October is one of the best months for the pond-hunter; the weather is cooler, the ponds have become filled with rain-water again, with plenty of food material in the shape of flagellate Infusoria, and the Crustaceans are on the decline. In this month the greatest variety in species of Rotifera is usually found, particularly of the smaller and rarer kinds, and not infrequently thirty to forty species may be obtained in one or two small ponds. As a general rule one cannot expect much variety when a few species are present in excessive abundance. The following is a list of forty-four species of rotifers actually collected on one occasion in three ponds on October 15, 1898, showing what may be looked for :

Floscularia regalis, ornata, cornuta, ambigua, edentata, and *annulata; Limnias annulatus, var. granulosus; Œcistes crystallinus; Philodina megalotrocha; Rotifer vulgaris; Synchæta tremula* and *oblonga; Asplanchna priodonta; Notops hyptopus; Polyarthra platyptera; Eosphora aurita; Furcularia longiseta, sterea,* and *forficula; Proales felis; Diglena biraphis; Mastigocerca rattus* and *bicornis; Cœlopus porcellus* and *tenuior: Rattulus bicornis; Diaschiza exigua; Distyla flexilis; Monostyla lunaris; Dinocharis pocillum; Stephanops lamellaris; Cathypna luna; Euchlanis oropha; Metopidia acuminata; Brachionus angularis* and *Bakeri; Pompholyx sulcata; Notholca labis* and *scapha; Anuræa aculeata, cochlearis, tecta, hypelasma,* and *stipitata.*

On the other hand, the various kinds of rotifers known as summer forms will now have disappeared. *Pedalion mirum* is such a form, which may occasionally still be seen during a warm October, but is then usually very scarce or absent.

November.

With the advent of November, pond-life all round becomes less abundant, and fewer species are to be met with. By degrees many of the water-plants die down, and the fauna is reduced to such forms as can subsist through the winter. Those animals which cannot do this, such as Polyzoa, Daphnia, some Rotifera, etc., have by this time produced so-called winter eggs or resting germs. The winter fauna, however, is much more numerous than is usually assumed. Among rotifers, several species of Synchæta—*S. pectinata, tremula,* and *oblonga*—seem to like the winter as well as the summer : *Asplanchna priodonta, Anuræa aculeata, Polyarthra platyptera, Rotifer vulgaris, Euchlanis deflexa, Triarthra longiseta, Brachionus angularis, Conochilus unicornis, Diglena forcipata, Diaschiza lacinulata* and *ramphigera, Dinocharis tetractis,* and others. Among the Infusoria, the Vorticella in particular seem to like the cold season, and a number of different species, and often large colonies can be found attached to submerged rootlets of trees growing near the edge of the water. Attached to the fine stems of Carchesium, Zoothamnium, and other stalked colonies of Vorticella, the very much more minute but beautiful colonies of Collared Monads, Codosiga, etc., are often found, and deserve a good look with the higher powers.

In canals and lakes where Cristatella has been abundant during the summer, their spiny stadoblasts may now be found liberated and often in large masses floating near the edge of the water which lies opposite to the direction of the prevailing wind. These should be collected and placed in a jar full of water with some Anacharis in the warm room at home, where they will hatch by the end of December or January, and the beautiful young Polyzoa can be seen emerging from their box-shaped prison.

December.

Severe weather in this country does not, as a rule, set in in December, and the lakes and ponds are not usually frozen over in the early part of the month. The winter fauna has now become more pronounced, but includes quite a number of Infusorians, Rotifers, and Crustaceans. The following species of rotifers have been collected in December in lakes and canals in and round London, some of them in great abundance : *Anuræa aculeata* and *cochlearis*, *Asplanchna Brightwellii* and *priodonta*, *Brachionus angularis*, *Diaschiza semiaperta*, *Euchlanis deflexa*, *Melicerta ringens*, *Œcistes crystallinus*, *Limnias ceratophylli*, *Floscularia cornuta*, *Synchæta pectinata* and *tremula*, *Conochilus unicornis*, *Rotifer vulgaris* and *macrurus*, *Polyarthra platyptera*, *Notholca scapha*, *Triarthra longiseta*. Of Crustaceans, *Diaptomus castor* and various Cyclops and their larvæ are abundant, whilst Water-fleas die down. A minute red flagellate Infusorian often seems to form the chief food material of the above lake fauna.

2. Collecting-Grounds near London.

A few of the principal collecting-grounds for pond-life in and near London may be mentioned. The nearest and most convenient available piece of water is the Grand Junction and Regent's Canal, which runs from east to west, on the northern side of London, from Victoria Park to Hanwell, and is readily approachable wherever access can be gained to the towing-path. Wimbledon Common and all the great parks have a lake, such as Victoria Park, Regent's Park, Hyde Park, Richmond Park, etc., which all afford good collecting-grounds. Smaller ponds are found in abundance in fields and commons in and beyond suburban London, and I need only mention a few such places : Epping Forest, Higham Park, Hadley Wood, Totteridge, Hampstead Heath, Ealing Common, Hampton Court, Putney Common.

3. Methods of Collecting, Preliminary Examination, and Keeping.

The fascinating study, under the microscope, of the living microscopic objects found in ponds, canals, and lakes, collectively known as ' pond-life,' requires, first of all, that you should catch your game. The object of this note, therefore, is to discuss those methods of collecting which, with a good many years' experience, have proved to me the most practical, efficient, and time-saving ; it is intended for the young naturalist or beginner who desires to make the personal acquaintance of these minute atoms of life, and thereby gain a better understanding of all living things.

A few pieces of apparatus are indispensable, and these are the following:

1. A Queketter's collecting-stick with ring-net and bottle, and cutting-hook.

2. A flat bottle.

3. A pocket magnifier.

4. A hand-bag with sundry wide-mouthed bottles.

The Collecting-Stick can be obtained from most opticians. It is a hollow walking-stick with an inner rod to increase its length when required, and provided with a screw at the end for the attachment of either ring-net, dipping-bottle, or hook.

The ring is a stout brass hoop, about 6 inches in diameter. The net, which is sewn on to the ring, is made cone-shaped,

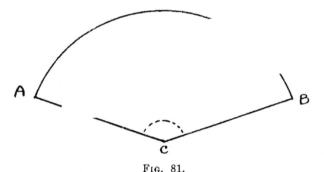

Fɪɢ. 81.

A, C = 9 in. ; angle at C = 140°.

about 6½ inches long, and at its apex is tied a small rimmed tube-bottle of clear glass, about 3 inches long by 1 inch wide. The material of the net should be either fine muslin, known as ' soft mull,' with meshes fine enough to prevent the Infusoria and Rotifera going through, and yet allowing the water to run out freely, or else a silk material known as ' Swiss bolting silk,' used by millers for sifting the various grades of flour, and obtainable from all mill furnishers ; No. 16 of this silk material has the required fineness.

The net is most important, and some care should be taken to have it properly made. Allowing a margin for the seam and for ewing round the ring, the shape and dimensions of the material for a 6-inch ring should be as represented in Fig. 81. This will give a net slightly larger than is required, but as the material is sure to shrink a little, it will be of the right size after having been used once or twice.

The cutting-hook is a curved knife which can also be screwed on to the collecting-stick, and is intended for cutting roots or water weeds which are otherwise out of reach.

The Flat Bottle can be obtained from opticians, well made, and the parts joined by fusing with fusible cement. When first invented by the late Mr. T. D. Hardy, it was made by cutting a LI-shaped piece out of a thick flat piece of india-rubber or similar material, 4 to 5 inches long, by 2 to $2\frac{1}{2}$ inches wide, and $\frac{5}{8}$ to $\frac{3}{4}$ inch thick ; a square of thin plate-glass of same size, cemented by means of Miller's caoutchouc cement on each side, completed the bottle. A thick piece of india-rubber is, however, so expensive that it is cheaper to buy the finished article. The flat bottle is used for searching over pond-weeds with the pocket lens at the side of the pond, or examining the water which has been collected and condensed with the net. In round bottles it is very difficult to see minute animals clearly, whilst a thin flat bottle allows the whole contents to be readily scrutinized with a pocket lens of considerable power, and one can at once determine whether it is worth while to take home a sample from that particular pond for further examination under the microscope.

The Pocket Magnifier best adapted for field-work is Zeiss's improved aplanatic lens, magnifying six diameters, which has a very large flat field, long focus, and perfect definition all over the field.

The various groups of plants and animals commonly designated as ' pond-life,' which inhabit fresh-water lakes, ponds, and ditches, consist of Algæ, Desmids, Rhizopoda, Infusoria, Sponges, Hydras, Rotifera, Polyzoa, Cladocera or Water-fleas, Copepods, Hydrachnida, Worms, and Insect larvæ. All these can be divided for the purpose of collecting into two groups—the free-swimming, and those that are usually attached to water-plants or submerged objects, and each of these groups must be captured in different ways.

All free-swimming or floating forms, which collectively are designated by the word ' plankton,' are best secured with the net. The net is passed through the water two or three or more times, and then held up ; the water will run out in half a minute and quite at the last the condensed animals will be seen entering the little bottle like a cloud, where they can be subjected to a preliminary examination. It is best, however, to empty the contents into the flat bottle, in which the examination with the pocket lens becomes very much easier, and most of the forms one is acquainted with can be recognized at a glance. In this way thousands of Algæ, Infusoria, Rotifera, Daphnia, etc., can be captured in a few minutes if the pond be a prolific one. Having thus ascertained that the dip contains some desirable forms, the water is poured into a large, wide-mouthed collecting-bottle, of which three to six should be carried in the bag. These bottles should be numbered; for it is often advantageous to keep the water of different ponds separate, so as to be able to know at home from which pond a particular creature has come. Ponds vary exceedingly as regards their contents in pond-life; a small pond may be very prolific, whilst another, possibly a larger piece of water only a few yards off, may contain hardly anything worth collecting. By trying all the different ponds, small and large, within reach of an afternoon's walk, one usually succeeds in obtaining a good gathering of free-swimming forms. The net quickly condenses a large volume of water, so that few species, even if present in small numbers only, will escape being captured. Several other methods of condensing pond-water have been devised, but the collecting-net with bottle attached is so simple and effective that we need not trouble about any other apparatus. It may be advisable to try the larger p. ... in various places, and both near the surface and also in deep water, as some plankton forms may have collected in one particular corner of the pond and be absent elsewhere ; this is often the case with *Volvox globator*. The use of a boat on larger lakes is very desirable when available. For rotifers and other active free-swimmers it is not desirable to disturb the mud at the bottom of the pond, but certain species of Cladocera, Hydrachnida, and insect larvæ can only be found at or near the bottom.

The group of attached forms of pond-life comprise such Infusoria as Carchesium, Epistylis, Zoothamnium, Stentor, etc.; Hydra; all Polyzoa and Sponges. In searching for these forms, a quantity of pond-weeds, or rootlets, are brought on shore with the cutting-hook, and selecting some likely-looking, fairly clean branches, but not the newest growth, one twig after another is placed in the flat bottle in clean water, where it can be examined from both sides with great ease, both with the naked eye and the pocket lens. The tree-like Vorticella colonies—Epistylis, Zoothamnium, Carchesium ; the trumpet-shaped Stentors ; the Crown Rotifer Stephanoceros ; the tubes of Melicerta and Limnias ; the various Polyzoa ; also Hydra and Sponges, and many others, can at once be seen when present, and in this way good branches can be selected and placed in a separate wide-mouthed collecting-bottle containing clean pond-water. A little experience will soon teach one which branches are likely to prove prolific. As a general rule one may say that old-looking, but still sound and green, branches are the best. The Water Milfoil (Myriophyllum) is one of the best water-plants to examine and collect on account of the ease with which its leaves can subsequently be placed under the microscope. Anacharis is more troublesome, but it is occasionally found covered with pond-life, and is an excellent weed for the aeration of aquaria.

The rootlets of reeds and of trees growing near the edge of the water should be examined for Sponges and Polyzoa, such as Lophopus, Plumatella, Fredericella, etc. In order to obtain some weeds growing near the middle of a pond or lake, a loaded three-pronged hook, attached to a line, may be used ; this is swung round, and may be thrown to a distance of 20 to 25 yards, where it sinks, and the weeds that are caught by the hooks are dragged on shore.

By these various means a good collection of pond organisms can readily be made after a little practice. Though the spring and autumn are perhaps the best seasons for collecting, pond-life is never absent, even in the winter under the ice.

Having thus filled some bottles with condensed water from various ponds, and placed some promising branches of water-plants in another bottle filled with uncondensed and clean pond-water, the ' bag ' is taken home. It is a great mistake, however, to overstock the bottles with weeds, as the plants in such crowded bottles may begin to decompose, killing most of the animals in a short time.

On reaching home, the first thing to do is to empty the collect-ing-bottle into larger vessels or small aquaria, in such a way that the captures may be critically examined, isolated, and, if found desirable, placed under the microscope. By far the best and most convenient way of doing this is to transfer the contents of each bottle into a small window aquarium, filling it up with tap-water. The weeds and rootlets that have been brought home are put in another window aquarium in clean pond-water.

These small window aquaria, with flat and parallel sides 6 to 8 inches long by 5 to 6 inches high, and only $1\frac{1}{4}$ inches wide inside, are the best nurseries for the microscope. The difficulty of seeing and capturing small objects in a large or ordinary round aquarium is very great, and the use of the pocket lens almost hopeless, whilst in these flat and narrow aquaria no object is out of reach of the lens, and the whole contents can be looked over without difficulty and in a very short time.

By placing the tank on a what-not at a convenient height before a window, or before a lamp at night, most of the free-swimming rotifers will collect against the glass nearest to the light, where they can be examined with the greatest ease and picked up with the pipette if desired. A disc of black cardboard placed some little distance behind produces a very good dark ground, against which the smallest visible specks stand out well.

The condensed pond-water is, of course, frequently so dirty with floating particles of débris that it is at first hardly possible to see through it; but after standing half an hour it will be found that most non-living particles will have fallen to the bottom, and after several hours the water will be quite clear and and every living creature will be readily seen.

During the summer months, when Daphnia and Cyclops are abundant, the net frequently collects these in such numbers that they become a nuisance. In order to separate them, when such is the case, I have adopted the plan of passing the water through a small sieve made of material with meshes sufficiently wide to allow the largest rotifers and Infusoria to go through, whilst keeping back most of the Cyclops and Water-fleas; the latter are then transferred to a separate tank to be examined by them-selves.

It is very desirable to examine the collected objects as soon as convenient, the same day if at all possible, and not later than the day after their capture, as many organisms soon die and disappear under the crowded and unnatural conditions in which they are kept in captivity. Rotifers can often, particularly in cool and cold weather, be kept for a week or fortnight, and some species, such as Melicerta, occasionally for months if food material in the shape of fresh pond-water can be provided. Failing pond-water, water from hay infusions, which mostly contain quantities of bacteria and minute Infusoria, may be added. The various species of Polyzoa and Sponges can also be kept alive a considerable time by feeding them in a similar way, but Hydras require a fare of Water-fleas if they are to thrive.

For keeping microscope life I have found no difference between large and small aquaria, but the small tanks are the more manageable; the great thing to be attended to is the proper aeration with water-plants, of which Anacharis, Fontenalis, and Valisneria are, perhaps, the best, and not to overstock the tank with either animal or vegetable life. The water need not be changed, but a little fresh pond-water should be added from time to time. Larger animals, such as small fish, water-beetles and snails must be excluded altogether from small tanks, and Polyzoa and Sponges must be kept therein in very moderate quantity and small colonies only.

In order to insure success it is essential to maintain a proper balance between the animal and the vegetable life, and also to supply fresh food frequently, for microscopical animals no more than the larger beasts can live long without food. To some extent, no doubt, they feed on each other, but in a small aquarium their hunting-ground is very limited and the game soon becomes scarce. Asplanchna can be seen under the microscope to feed on Anuræa, Brachionus, Polyarthra, Triarthra, and other rotifers when it can catch them, and their shells and remains are frequently found in Asplanchna's stomach.

On the whole, the best plan is to go out and collect a fresh supply from time to time, and as often as may be convenient. I may mention that at the middle of January I had many thousands of rotifers in a tank which I collected two days before in the Grand Junction Canal, near Westbourne Park Station. The canal was covered with blocks of ice, and the time spent near the water did not exceed ten minutes, during which I filled a large bottle with water condensed by means of the ring-net.

Everyone who has worked at pond-life will have experienced how awkward it is to examine with a pocket lens, and at the same time attempt to pick out a particular animal in order to place it under the microscope. In order to have both hands free for this operation, and to keep the lens fixed to a particular spot, I devised some years ago a small aquarium microscope (Fig. 82), which is simply a flat metal arm, jointed in such a

Fig. 82.

way that it allows the lens to be moved all over the surface, but in one plane only, parallel to the side of the window aquarium, whilst the lens is focussed by a small rack and pinion on the left. The whole apparatus is screwed to a small wooden stand, on which the tank is placed. The lens used is Zeiss's aplanatic combination × 6 diameters, which has working distance enough to focus right through the tank, and sufficient amplification to enable one to recognize most rotifers, Infusoria, etc., and anything uncommon or new can at once be detected and secured. Moving objects can readily be followed with the lens, and pond-weeds can be searched for anything that may be growing on them, whilst the lens remains fixed in any position it may be placed. I have had this tank microscope in constant use for over twelve years, and can recommend it as thoroughly practical, efficient, and time-saving.

4. Apparatus for Microscopic Examination.

I propose now to describe those methods which long experience has proved to be the most practical in the examination of living objects under the microscope.

Fig. 83 is a photograph of various apparatus used for this purpose, consisting of troughs, pipettes, live-box, and compressor.

After capturing a miscellaneous collection of pond-life and transferring it to a window aquarium placed in front of a window,

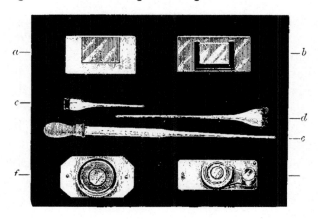

FIG. 83.

as previously explained, it will be desirable first of all to place some of it under a low power of the microscope—say a 2-inch or 1½-inch objective—in order to obtain a better general view of the various animals. The free-swimming forms will mostly have collected on the light side of the aquarium, and can there be picked up quite clean and in vast numbers, sometimes with the pipette *e*, and transferred to a square trough *a* or *b*, and placed under the microscope, where the contents can readily be illuminated from below, both with transmitted light and under dark ground. I prefer to use dark-ground illumination with low powers when searching over the contents of a trough, and when studying the shape, mode of swimming, ways of feeding and living of Polyzoa, Rotifera, and Infusoria. Moreover, the animals scattered through the trough will soon collect in the spot of light of the condenser, and then the whole field of view will often be a mass of moving, dancing, tumbling, sparkling life.

The trough a, 3 inches by $1\frac{1}{2}$ inches and $\frac{5}{16}$ inch thick, is the form I mostly use; it stands upright on the table, is reversible, and can be handled without greasing the well part of the glass. The sides are cemented in the fire by means of a fusible glass cement, and thus the trough is, and remains, watertight. The trough b is also a useful type, but it is not reversible, will not stand by itself on the table, and, being cemented with gold size or marine glue, is liable to leak. The troughs usually sold are semicircular in shape, a very bad type, because, in addition to the above defects, the least amount of tilting on the stage will cause the water to run out over the edge. Thicker troughs are objectionable because the sub-stage condenser cannot work through them, and the animals cannot be properly illuminated, though sometimes such troughs may be required by the size and nature of the object.

A few words on pipettes will not be out of place here. The old-fashioned way of using the finger on a straight or curved glass tube to capture pond-life is so unsatisfactory that I have been driven to invent new pipettes for more precise and exact work. Fig. 83, c, d, and e, represent the pipettes in constant use; e is a glass tube about $\frac{5}{16}$ inch in diameter and 8 inches long, which tapers from the middle to a point more or less fine, according to the size of the animals one wishes to capture. Over the wide end is placed an india-rubber teat, by means of which any single specimen, or scores of animals, can be sucked up with the least quantity of water; d is another type of pipette, having a still finer action; it is 6 inches long, funnel-shaped at one end, and tapering gradually from the funnel to a fine point; the funnel is $\frac{3}{4}$ inch wide, and covered with an india-rubber membrane; c is a similar, but smaller and finer pipette, $3\frac{1}{2}$ to 4 inches long, for picking up small rotifers in a fraction of a drop of water under the dissecting microscope. The slightest touch on the membrane is sufficient to expel or bring in the water, so that one has complete control over the amount of water that is taken up, and there is much less risk of losing the animal one wishes to transfer to the compressor.

The old-fashioned live-box with raised tablet, still largely sold with microscopes, is quite useless for pond-life, for the simple reason that the objects cannot be properly illuminated with the sub-stage condenser. This consideration led me long ago to design the live-box *f*, in which the glass tablet is fixed flush with the brass plate, and is of small size, thus leaving a wide ring all round. This arrangement allows all objects on the tablet to be perfectly illuminated from below by the achromatic condenser, both with transmitted light and under dark ground, and at the same time they can be reached and followed from above with both low and high powers and oil immersion lenses, to the very edge of the tablet, and wherever they may wander. For more exact work, when it is desired to hold a single rotifer between the two glasses, and prevent its wandering about, I have devised the compressor *g*, in which the pressure and the thickness of film of water can be accurately regulated by a screw acting against a spiral spring. At the same time, water, or reagents, can be added, if desired, without raising the cover. When properly and well made, this compressor works exceedingly well, and I have had it in constant use for years, but some makers, unfortunately, have introduced variations and so-called 'improvements' which just take away some of the essential and useful points. The semicircular thin cover-glass must be cemented to the under side of the brass ring with a little gold size, so as to be quite firm and rigid, otherwise its action becomes uncertain, and very small objects cannot be held fast, or else are suddenly crushed.

Some more simple apparatus and devices may be mentioned for cases where no live-box or compressor is at hand. An excavated glass slide makes a fair live-cage ; a drop of water containing the animals is placed in the cavity so as to just fill it, and no more ; another drop of clean water is placed by the side of the cavity, and a clean thin cover-glass is lowered on to that second drop ; then, by means of a needle, the cover is slowly pushed across the cavity, which can thus be covered without enclosing an air-bubble ; the superfluous water is taken up by blotting-paper, the cover-glass being held in position by capillary attraction. This forms a good slide for low and medium powers, but not for high powers. Another good temporary slide can be made by placing three small fragments of No. 1 thin cover-glass near the middle of a glass slip in form of a triangle ; the drop of water containing the animals is placed in the centre, and a clean thin cover-glass is lowered on to the drop so as to rest on the three glass fragments, which prevent the animals getting crushed. If there be too much water it can be removed with blotting-paper. Low and high powers can be used on this slide as far as the movements of the animals will permit, but not oil immersion lenses.

Having thus mentioned some essential and necessary apparatus, I will close with a few remarks on the examination of living pond-life. The free-swimming organisms, including such forms as *Volvox globator*, collect on the light side of the window aquarium, and can there be picked up in small or large numbers, and quite clean, with the large pipette, and placed in the trough ; or any particular species can be selected with the aid of the tank micro-scope, and taken up with the smaller pipette, and transferred to the live-box or compressor in a single drop of clean water, both hands being free for this operation.

The fixed forms, such as Polyzoa, Stephanoceros, Melicerta, Floscules, etc., amongst Rotifera, and Stentor, Carchesium, Zoothamnium, etc., amongst Infusoria, require a little manage-ment. If simply placed in a trough, these are often obscured or incapable of being properly illuminated by being too crowded, or by part of the weed over- or under-lying the objects, and also by floating particles in the water. The best result is obtained by trimming—that is, by cutting off a very small piece of weed or leaf on which the animal is attached—in a watch-glass under the dissecting microscope, if necessary—and then transferring it with the pipette to the compressor into a drop of clean water ; it can then be arranged with a needle or bristle as may be desired, and after lowering the cover-glass, fixed and held fast, at the same time giving the animal perfect freedom to expand. In this position the animals can be reached with the achromatic condenser from below for transmitted light and dark-ground illumination, and also with low and high powers, and even oil immersion objectives from above.

For pond-life work the Wenham binocular is decidedly to be preferred to the monocular microscope. It is less tiring to the eyes to look with both eyes and without strain, and the stereoscopic image gives a very much better idea of the true shape of the animals, though the images are not quite so sharp as with the monocular tube; but this binocular form can immediately be changed into a monocular for high-power work, or whenever desired, by pushing the small prism out of the way.

The binocular is to be used only with the low powers up to the $\frac{2}{3}$-inch objective; with higher powers the stereoscopic effect is lost, because the depth of focus, or the plane of distinct vision, is then exceedingly small, and becomes more and more a mere optical section of the object.

A mechanical stage is hardly necessary; for ordinary work a well-made sliding stage or bar is preferable, and should be provided. Stage-clips, of which opticians are so fond, are abominations, and should be consigned to the dust-bin.

Of illuminating apparatus, the Abbe form of sub-stage condenser, achromatic if possible, is the only one that is really useful for all powers, and that need be considered both for transmitted light and for dark-ground illumination. It should be provided with an iris diaphragm and an arm carrying a central stop; it completely replaces all the older sub-stage apparatus—condenser, spot lens, paraboloid, etc. The bull's-eye stand condenser, however, is necessary to render parallel the rays of the lamp-flame, but it should be mounted on the lamp, and move about with the lamp, and so as to project an enlarged image of the edge of the flame on to the flat mirror of the microscope for dark-ground illumination with low powers.

All apparatus used in the examination of pond-life—troughs, live-boxes, compressors, and pipettes—should always be carefully cleaned and dried immediately after use, and in no case should the water be allowed to evaporate in them. Much trouble will be saved by the observation of this rule, and the apparatus will always be ready for use.

5. Preserving and Mounting.

There are few observers of pond-life who have not felt a keen desire to preserve and keep these small highly organized sparks of life instead of letting them die and disappear in a few days. For a close study of this group, well-preserved type

17

specimens are of the greatest possible assistance and importance, and if such had existed formerly much confusion and inexactitude in their description and classification would have been avoided, particularly in the giving of three or four different names to the same species, which causes so much trouble to the student.

The total absence of type specimens of rotifers to refer to when required, originally led to an attempt on the part of Mr. C. F. Rousselet to produce them, and it is now over ten years since the first successful experiments at preserving them in a fully extended and natural state were made. His method, although so simple now, took fully three years to work out until the right and most suitable narcotic, fixing agent, anc preserving fluid were found. By the use of suitable fixing agents not only the external shape of rotifers can be preserved, but also all the internal structure, to the minutest anatomical details, such as the striated muscle fibres, nerve threads, vibratile tags or flame cells, sense hairs, cilia, etc., and frequently important details can be more readily observed than in the living animal.

Narcotizing.—As is well known, no killing agent is sufficiently rapid to prevent the complete retraction of rotifers, and few other animals can contract into such a shapeless mass when we attempt to kill them by ordinary means, such as poisons, alcohol, heat, etc. It is, therefore, necessary to use first ·a suitable narcotic, which has been discovered in hydrochlorate of cocaine. As a result of many trials, the best solution for most rotifers has been found to be the following mixture :

2 per cent. solution of hydrochlorate of cocaine, 3 parts ; alcohol (or methylated spirit), 1 part ; water, 6 parts.

Another narcotic which is also very suitable for rotifers is a 1 per cent. watery solution of hydrochloride of eucaine, recommended by Mr. G. T. Harris, for Infusoria and other animals. These narcotics, even so dilute, are not to be used pure, as they would cause the rotifers to contract at once and not expand again. The principle to be followed throughout is to use the narcotic so weak that the animals will not mind it at first, but continue to expand or swim about freely. After a short time its effect will make itself felt on their nervous system, and then some more of the narcotic may be added, until complete narcotization is produced, or until the animals can be killed without contractings.

But before the operation of narcotizing is begun, it is very necessary to isolate the rotifers in perfectly clean water. The best way is to pick them up under a dissecting microscope by means of a very finely drawn-out pipette, having a funnel-shaped enlargement at the other end, which is covered with an elastic membrane. This pipette forms a most delicate siphon, by means of which any selected rotifer can readily be taken up with the least quantity of water, and transferred to another trough or watch-glass full of clean water. This preliminary precaution is necessary, because particles of dirt in the water readily attach themselves to the cilia of dead rotifers, rendering them unsightly under the microscope. Another advisable precaution is to separate the different species, because most species require a slightly different treatment, and because the small species too readily adhere to the cilia of the large species.

Having then isolated a number of free-swimming rotifers in a watch-glass half full of perfectly clean water, one drop of one of the above narcotics is added and well mixed. After five or ten minutes, if the animals continue to swim about freely, another drop is added, and so on until the effect of the narcotic becomes visible, and until the motion of the cilia or the movements of the animals slacken or almost cease, when they are ready for killing. The effect of the narcotic varies very much with different species; some are most sensitive to it, whilst others can stand a considerable quantity for a long time.

Killing and Fixing.—Some practice and patience are certainly required to find out the right time to kill the different species; no general rule can be given, as the time may vary from fifteen minutes to several hours. It is very essential, however, that the rotifers be still living when the killing fluid is added to prevent post-mortem changes in the tissues, which begin at once on the death of the animals.

For killing and fixing several fluids are suitable—namely, $\frac{1}{4}$ per cent. osmic acid, or Flemming's chromo-aceto-osmic fluid, or Hermann's platino-aceto-osmic mixture. On the whole, I now prefer the last-named, which gives a finer fixation of the cellular elements of the tissues and does not stain them so much. It may be explained that the term 'fixing' implies rapid killing and at the same time hardening of the tissues to such an extent as to render them unalterable by washing and subsequent treatment with preserving fluids. Proper fixation is very essential, as no good preservation can be obtained without it.

When the rotifers are narcotized and ready for killing, a single drop of one of the above fixatives is added, and mixed with the water in the watch-glass. A few minutes is sufficient for fixing small creatures like these, and then they must be removed again by means of the pipette to several changes of clean water to get rid of the acid, otherwise they will become more or less blackened. When dealing with marine rotifers, sea-water must be used for washing out, for the difference in density between fresh and sea water is sufficient to cause swelling by osmosis, and the consequent spoiling of the specimen. After thoroughly washing, the rotifers are transferred to a preserving fluid, the density of which does not materially differ from that of water. The best preserving fluid found so far is a $2\frac{1}{2}$ per cent. solution of formalin, which is made by mixing $2\frac{1}{2}$ c.c. of the commercial 40 per cent. formaldehyde with $37\frac{1}{2}$ c.c. of water, and then filtering.

The above are general directions according to which the great majority of rotifers can be preserved. When under the narcotic, the animals must be watched until it is seen that they can swim but feebly, when, as a rule, they will be ready for killing. If they contract and do not expand again, it is a proof that the narcotic used is too strong, and it must be further diluted. The whole method undoubtedly requires great care, and is a delicate operation, which must be performed under some kind of dissecting microscope, but by following the directions here given, and with some perseverance, anyone can learn to prepare a large number of species of rotifers. I would advise that a beginning should be made with some such forms as Brachionus, Anuræa, Synchæta, Asplanchna, Hydatina, Triarthra, and Polyarthra, which are easy, and, moreover, occur, and can, as a rule, be collected in large numbers. A few genera, however, are exceptionally difficult. These are Stephanoceros, Floscules,

Philodina, Rotifera, and Adineta, and it will be better to leave these until considerable experience in dealing with the others has been acquired.

It will have been noticed that the rotifers must always remain submerged in a watery fluid, and be transferred in a drop by means of the pipette. Fluids of lesser density than water, such as alcohol, as well as fluids of greater density, such as glycerine, are unsuitable because they set up strong diffusion-currents by osmosis, which cause the animals either to swell or to shrivel up completely.

Some species of rotifers, such as Triarthra, Polyarthra, Pedalion, Mastigocerca, etc., have an outer surface which is strongly water-repellent, and when these come in contact with the surface film of the fluid even for an instant it is most difficult to submerge them again, and, as a rule, they are lost and spoiled.

Having then successfully narcotized, killed, and fixed the rotifers fully extended, and finally transferred them into $2\frac{1}{2}$ per cent. formalin, the animals may be kept in little bottles, or mounted in the same fluid on micro-slides, either in excavated cells or shallow cement cells.

Mounting.—To mount on a slide, place a drop of the formalin solution in the cell, then transfer the prepared rotifers into this drop with the pipette, and examine under the dissecting microscope to see that no particle of foreign matter has been introduced. Then place another drop of the fluid on the slide by the side of the cell, lower the cleaned cover-glass on that drop, and push the cover cautiously and gradually over the cavity. The superabundant fluid is removed with blotting-paper, and the slide closed by tipping damar-gold size cement all round the edge with a fine brush.

The permanent closing of these cells has been a matter of very considerable difficulty. As the result of the experience gained, it is recommended that the cells be closed first with a coat of a varnish consisting of two-thirds damar in benzole and one-third gold size, then two coats of pure shellac dissolved in alcohol, and finally four to six coats of pure gold size. Each layer of cement must be allowed to dry thoroughly well; three days for each layer is not too long.

By the method described above, Mr. Rousselet has in the course of the last ten years made a collection of over 500 slides containing nearly 300 different species of rotifers, probably the only collection of the kind in existence, which is of the greatest use for the identification of species and for the general study of this interesting class.

Entomostraca should be narcotized with the same solution as used by Mr. Rousselet for Rotifera, then killed with a $\frac{1}{4}$ per cent. solution of osmic acid, and mounted in a $2\frac{1}{2}$ per cent. solution of formalin.

CHAPTER XXIII

THE COLLECTION, EXAMINATION, AND PRESERVATION OF MITES FOUND IN FRESH WATER

By C. D. SOAR, F.L.S., F.R.M.S.

ANYONE with a love for natural history wishing for a hobby for his spare time, would find the study of fresh-water mites (Hydrachnidæ) an extremely interesting one. For variety and beauty in colour, and for differences in form and structure, they are not to be surpassed by any other organisms found in fresh water. Wherever there is a pond, ditch, or stream, the collector is nearly sure of being rewarded for his search by finding one or more species of these interesting creatures. They are easily caught, and can be seen with the naked eye ; they are, however, very seldom recognized without the aid of the microscope. They can be kept alive for a considerable period at home, and are easily preserved when killed.

At present the life-history of these little creatures is so imperfectly known that there is wide scope for an observant naturalist. Although the life-histories of some species have been fairly investigated, the number of such is very limited compared with the species known, and the variety of species which have been recorded in Great Britain are behind the recorded collections of Germany and elsewhere.

These creatures are caught in three distinct stages—the larval, the nymph, and the imago. In the larval stage they are very small, and only have six legs. When they first emerge from the egg they are free-swimming, but they soon become attached as parasites to some other form of pond-life. They will often be

NOTE.—The figures in this chapter are from the *Journal of the Quekett Microscopical Club*, by permission.

found hanging like small red pear-shaped appendages on a great number of aquatic insects. The six legs they started life with disappear after they have become firmly attached by their mouth-organs to their host, and they spend the remainder of this period of their existence without any.

This stage is succeeded by the nymph ; the little creatures are then much larger and have eight legs. During this term of their existence they are free-swimming, and can be caught in the net in numbers, but it is impossible to distinguish the sexes.

In the last stage—the adult or imago—all the structure and form are present, but many may be taken that are not fully developed. In the majority of species, the male can be dis-

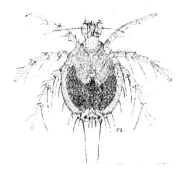

FIG. 84.—LARVA OF PIONA LONGIPALPIS. (KREN.)

tinguished from the female and the specific differences recognized ; but there are some in which the sexes are so much alike that it is almost impossible to tell one from the other. In others, again, the sexes are so different—as, for instance, in the Arrhenuri—that one would be disinclined to think they could be of the same species.

The three figures are intended to convey to the beginner the three stages mentioned. Fig. 84 is the larva of *Piona longi-palpis*. Fig. 85 the nymph and adult of *Hydrachna globosa* (Geer), showing the ventral surface and the epimeral plates to which the eight legs are attached. Fig. 86 is the larva of an Hydrachnid, parasitic on *Dytiscus marginalis* and *Nepa cinerea*.

There is another point in the adult stage to which it will be well to draw attention. When the mite has first made its appearance from the inert period it spends between the nymph

and adult stage, the hard and chitinous parts appear to be
nearly fully developed, but the soft parts are not so. The body
often appears very small, while the palpi, legs, and epimera, etc.,
are very large in proportion ; it is also very poor in colour. It
would be well to ascertain that the mites are quite developed
before making drawings and taking measurements. In my
gnorance, when I first began the study of water-mites, I had
to discard a number of drawings I had made of different speci-

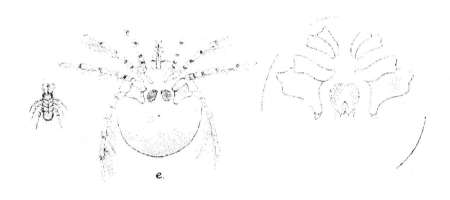

FIG. 85.—LARVA, NYMPH, AND ADULT HYDRACHNA GLOBOSA. (GEER.)

mens because they afterwards proved to be only different stages
of growth of the same species of mite.

For collecting mites there is no better apparatus than the
usual collecting-stick used by pond-hunters, having a metal
ring attachable at its end which carries a cone-shaped net made
of silk or muslin, with a glass tube at the bottom. The
advantage of the tube is that the contents can be examined
with an ordinary pocket lens at any moment to ascertain if
anything has been secured worth preserving.

It is advisable to carry as many bottles as the number of ponds
that are likely to be visited ; careful record should be kept of the

exact locality where each mite is found, with the date of capture, and this cannot be done if all the specimens are carried home in one bottle.

The most convenient way of carrying collecting-bottles is by

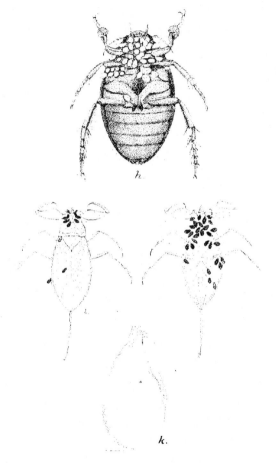

FIG. 86.—SHOWING LARVÆ OF AN HYDRACHNID PARASITE ON DYTISCUS MARGINALIS AND NEPA CINEREA.

h, *Dytiscus marginalis*, showing parasites on ventral surface and leg; *i*, dorsal surface of *Nepa cinerea* ; *·*, ventral surface of ditto ; *k*, larval parasite detached.

sewing two strips of thick cloth together with loops of the required size in the same manner as a cartridge bandolier. Such a device can be rolled and stood at the bottom of a bag, and obviates the chance of the bottles breaking by contact.

During the summer months it will generally be found that the most successful captures are made near the edges and in shallow parts of ponds; in the winter-time the mites get into deeper water. Some mites are to be found only on the mud at the bottom of ponds, others on the leaves and stems of water-plants. In collecting, therefore, it is necessary to let the edge of the net just skim over the surface of the mud and sand, and up and down the stalks and stems of likely plants.

The under surfaces of leaves should also be scraped with the edge of the net. Anacharis is a very favourite plant of water-mites, and wherever this is found it is almost certain that mites will be secured.

In addition to the free-swimming mites, there are a large number of parasitic forms, and it is as well to examine all forms of insect-life before discarding material. Fresh-water mussels, in particular, also the large water-snails and water-beetles, are specially to be recommended for examination, and once more let me emphasize that if anything is found, notes should be made of dates, places, and general details of the captures.

On reaching home the contents of the bottle should be emptied into a porcelain dish such as photographers use, when it will be noticed that the mites generally swim in the corners or along the sides, and can then be removed with a pipette to a large tube filled with clean water in which some Anacharis is placed. This latter will keep the water clean and fresh for a considerable time.

Experience will dictate which species can safely be kept together, a matter in which some discrimination is required, because some varieties prey on others—such, for instance, as Limnesia on Eulais.

Undoubtedly the best plan is to proceed with the examination at once, because a great part of the brilliancy of colouring is lost in a short time, and the mites are much more lively when freshly caught than subsequently. I have, however, kept mites alive in a tube 4 inches by 1 inch, by adding fresh water to replace that evaporated, for a period of twelve months.

The best method of examination is to place the mite on a 3-inch by 1-inch glass slip, turning the specimen on the ventral or dorsal side as may be required, and having every part extended. A cover-glass is then laid over the specimen, and sufficient clean water is allowed to flow between the cover-glass and the slip to fill the intervening space. The specimen may move its limbs and palpi for a short time, but soon becomes quite passive, the weight of the cover-glass being sufficient to retain the body of the mite in position. The slip is then laid on a piece of white card on the stage of the microscope, and illuminated by reflected light; a 1½-inch objective will usually be found the most suitable.

The advantage of this arrangement is that the specimen can be reversed, and both sides examined, and by having an aperture in the cardboard, a further examination may be made by transmitted light. In this latter condition the hairs and claws can be seen very distinctly, particularly if the light be thrown a little obliquely. After examination the specimens can be returned to the tube, and are usually none the worse.

To preserve the specimens they should be placed in the following solution :

10 parts glycerine, 10 parts distilled water, 3 parts citric acid, 3 parts pure spirit.

They can be placed in the solution alive, and although at first the limbs will be contracted, they subsequently retract. It also preserves the colours of hard-skinned mites fairly well.

If at any time it is desired to make a mounted preparation of any mites preserved in this way, they can be transferred to cells containing the same solution. If required for balsam mounts, the glycerine can be removed by repeated soaking in absolute alcohol, subsequently passing them through clove oil.

It will be found that balsam-mounted specimens will have a tendency to vaporize; this can be obviated by making a small hole in the body of the mite in a position which is of no consequence, and thus allowing the balsam to penetrate. I think the soft-skinned mites mount best in glycerine solution; I do not mount in this medium myself, but have some beautiful preparations by Mr. Taverner, in which the construction is shown to the best advantage. They have been in my possession for some time, and show no signs of deterioration.

Should any readers take up the study of these beautiful creatures, dates of collecting, localities where discovered, and particulars of anything they may have observed new in the life-history, particularly varieties of colouring, should be carefully kept, together with, if possible, drawings. There is one mite, *Piona rufa Koch*, which has been found in England in three distinct, bright, and beautiful colours—viz., red, green, and brown.

The two best textbooks on fresh-water mites are in German— 'Deutschlands Hydrachniden,' by Dr. R. Piersig, rather an expensive work, with about 500 pages of letterpress and 51 plates; and a number of 'Das Tierreich' on the Hydrachnidæ, by Dr. R. Piersig, Berlin, 1901. This contains the account of every known species up to date of publication.

CHAPTER XXIV

COLLECTING AND PREPARING FORAMINIFERA *

By ARTHUR EARLAND, F.R.M.S.

THE foraminifera, in spite of their beauty, the important part which they have played in the building up of our earth, and the many interesting features of their life-history, have not met with so much favour among microscopists as many groups of far less importance. This comparative neglect is largely due to mistaken ideas as to the difficulty of obtaining and preparing suitable material, and it is proposed to show, so far as possible within brief limits, that the collection of material is within the reach of every visitor to the seaside, and that the subsequent preparation presents no unusual difficulty to the microscopist.

The chief sources from which foraminifera may be obtained are :

1. Dredged material, including anchor muds and sands.

2. Shore gatherings made between tide marks.

3. Sands, clays, and limestones of various geological ages, especially from cretaceous and tertiary deposits.

As probably very few readers will have the opportunity of dredging for material, and as anchor muds, which often contain an abundance of shallow-water forms, are rarely obtainable, owing to the strange reluctance of seamen to lend themselves to the collection of scientific material, it is not proposed to enter at any length into the methods of collection by means of the dredge. The method of preparation for materials of this class is essentially the same as that for shore gatherings.

The ordinary naturalist's dredge can be successfully used for

* Reprinted from *Knowledge* (with additions).

the purpose of collecting foraminiferous material from the sea-bottom, but if the dredge is of the usual type, with the bag made of rope-net, it will be necessary to insert a canvas lining to the lower end of the bag, in order to insure the retention of some of the finer sand and mud. The size of the canvas bag must be governed by the strength of the dredge and the power of the lifting gear. A large deep-sea dredge such as is used on scientific cruisers will have a mouth 4 or 5 feet in width and a bag 6 feet long. Such a dredge will on a soft bottom fill up in a quarter of an hour or less, and as it weighs a ton or more, the lifting gear must be correspondingly powerful. When the dredge is operated and lifted entirely by hand, quite a small bag in the end of the small naturalist's dredge will be as much as can be managed.

After the dredge has been emptied on board, the material is usually washed in a large tub through a series of sieves from $\frac{1}{2}$ to $\frac{1}{8}$ inch mesh, in order to separate the mollusca and other large organisms. A supply of the mud *as dredged* should first be set aside in a canvas bag and labelled 'Original deposit.' This would be required for any quantitative analysis of the organisms contained in the mud, as many organisms would be lost during the washing process.

After selecting the softer organisms from the residue left on the various sieves for preservation in alcohol or formalin, the residue can be transferred to separate canvas bags and labelled with particulars of the sieve from which it was obtained. Large species of foraminifera, and especially arenaceous types, will be found retained on the $\frac{1}{8}$-inch sieve. The coarser sieves will as a rule contain only stones and molluscan or echinoderm fragments, but these will often be found to be covered with parasitic foraminifera.

The tub in which the material has been washed will now be found to contain a bottom deposit of mud and sand and a large quantity of muddy water. As a rule, this muddy water contains an abundance of the smaller species of foraminifera in suspension, and as it takes a long time for them to settle down (in rough weather the motion of the ship will keep them constantly in suspension), the muddy water should be baled off and strained through a fine silk net, such as a tow-net. By this means many species may be obtained in abundance which would otherwise escape observation. After the water has drained away through the net, the fine mud may be preserved in a bag, or, preferably, in alcohol. Formalin, having an acid reaction, should never be used for the preservation of foraminifera.

The bottom deposit of mud from the tub should then be preserved in canvas bags for preparation ashore. If the bags are thoroughly, but slowly, dried over the engine-room, and stored in a dry place, the material can be preserved for many years uncleaned and without deterioration, although it is better and easier to clean it as soon after collection as possible.

The apparatus required by the shore-collector is of the simplest character, and consists of a scraper for removing the surface film of sand, which alone contains foraminifera, a spoon for scraping material from ripple marks and depressions, and a metal box, or canvas bag, to contain the gathering. The best scraper is a thin plate of celloidin (about the thickness of a visiting card), such as a ' photographic ' film, as the thinness and flexibility of this material enables the collector to make his scraping with less admixture of sand than is possible with the glass or metal slip usually recommended for use.

Thus equipped, the collector sallies forth between the tides. Probably everyone has noticed when at the seaside the white lines which run along the sands parallel with the retreating tide. A pocket lens shows that the white material consists largely of the minute shells of foraminifera, of which some are of a lustrous white colour, due to the comparative abundance of the Miliolidæ —a family of common occurrence in shore gatherings, characterized by opaque shells of a milky white or ' porcellanous ' texture—while others are more or less glassy and transparent. These ' hyaline ' forms are much less noticeable to the naked eye. They are mixed in varying proportions with fragments of shell substance—ostracode shells, cinders, and the lighter débris of the shore—and their presence in these lines is due to the separating action of the water, which on a smaller scale we shall later on employ in the cleaning of our collected material. The rocking action of the wave on the extreme edge of the ebbing tide keeps these shells and fragments of light specific gravity in suspension until after the heavier sand-grains have subsided, and so they are left behind in the ripple marks and depressions of the sand. Sometimes a local eddy of the tide, produced by the neighbourhood of a projecting rock, or of groins and piers, causes the material to be gathered together in large quantities, which show as extensive white patches on the sand, and prove a real gold mine to the collector, who will then obtain more material in half an hour than he could gather in several days from the ripple marks.

The collector must not conclude that there are no foraminifera present because there are no white patches to be seen, but, remembering the way in which these patches are formed of the lighter débris of the shore, must look for foraminifera wherever he observes that such débris has been deposited.

On every coast, at intervals of varying distance, there are spots which appear to be the foci of the local tides and currents, and here the material will be found in the greatest abundance. These points will soon be discovered, and may be worked at every tide, but they vary continually with the set of the tide and wind, so that a spot which has proved rich may be quite bare the next year. Thus, in October, 1896, Bognor—always a rich collecting-ground—had its richest point to the west of the pier; while in September, 1901, there was very little material obtainable except at Felpham, two miles to the east, where the beach was thick with débris.

Having found the material, the collection is quite an easy matter. With the celluloid scraper at an angle of 60°, the thin surface film of foraminifera and débris is easily scraped into a heap, and transferred to the box or bag. Great care must be exercised not to dig down into the sand, for nothing but a heavy bag will result from this, the foraminifera being confined to the surface layer. The material thus collected may be either cleaned at once, or, after being slowly dried—avoiding great heat—may be packed away in bottles for a more convenient period.

The apparatus required for the cleaning and preparation of the dried material is simple and inexpensive, and, if desired, much of it may be easily improvised. The most necessary articles are a photographic developing-dish of china, quarter- or half-plate size according to fancy, sieves of different sizes and materials according to the collector's pocket, a cylindrical glass jar with a lip, and without any neck or constriction at the top, or a large plain glass jug, and a retort stand or tripod, made of an iron ring riveted on three legs.

The sieves can be made by any coppersmith, and it is very convenient to have a series of varying degrees of coarseness; but for the beginner, two sieves of 40 and 120 meshes to the inch respectively will be sufficient. The writer's sieves are of copper, 4 inches high, 4 inches diameter at top, sloping to 3 inches diameter at the bottom. A smaller size, made of telescope-tubing $1\frac{1}{2}$ inches in diameter and 1 inch deep, is very useful for washing small gatherings. The writer also uses Mr. Heron-Allen's method of serial sieves fitted one within another. But great care is required in the use of this method, as the lowest sieves, having the finest mesh, speedily become choked with material, and then overflow, causing the loss of specimens. Zinc, which is cheaper than copper, can be used for the sieves.

The wire gauze, which can be obtained from any large iron-monger, varies in price according to the number of meshes to the inch, ranging from a few pence per square foot to four shillings for the finest obtainable, which has 120 meshes to the inch, the diameter of each aperture being about $\frac{1}{200}$ inch. If a finer sieve than this is required, as it sometimes may be, the size of the aperture may be reduced by silver-plating the gauze, or, preferably, by the use of silk bolting cloth, which may be obtained up to 200 meshes to the inch. The wire gauze must be strained tightly over the sieve and soldered neatly to the edge, so that there is no ledge of solder inside to retain un-washed material. If silk is used, a sieve must be made without a bottom, and having a turned-back edge at the lower end, so that the silk may be strained across and secured with string or a rubber band. Very useful and effective sieves can be made out of brass tube 3 inches diameter, having a detachable clamp-ing collar operated by a screw. The silk is strained over the tube and held in position by the collar. They are much easier to clean than the usual sieve, but are expensive to make. The most useful sizes for a series of sieves are, in my opinion, 12, 20, 40, 80, 120, and 150 (silk) meshes to the inch.

Before cleaning the shore material, it must be slowly and thoroughly dried. It should then be passed through the twelve-mesh sieve to remove all the coarse débris, stones, shells, cinders, etc. Few, if any, of the British shore species, except parasitic forms, will be found in this coarse residuum, but it should be looked over with a pocket lens for these, or for abnormally large specimens. In some dredged materials and in tropical gatherings, however, this coarse residuum will be found to be full of foraminifera.

The material which has passed through the twelve-mesh sieve consists of foraminifera mixed with other light débris and a considerable quantity of sand, and the collector must now proceed to eliminate the whole, or nearly the whole, of the sand, and as much as possible of the other débris, by means of two operations—'floating' and 'rocking.' If the quantity of material to be operated upon is small, it may be treated off-hand, but if there is much, it is well to sift it out into varying degrees of fineness by passing it through a series of sieves. This will simplify the floating operations by insuring that the particles are approximately of similar weight.

The floating operations must be performed at a sink, and, if possible, in daylight, the process being more uncertain by artificial light. The finest sieve (120 wire or 150 silk) is thoroughly wetted and rested on the tripod. The glass jar is then filled with water nearly to the brim, and a few spoonfuls of sand slowly poured into it. If the material is coarse the sand sinks instantly, and in the course of a few seconds most of the foraminifera follow suit. By holding the jar to the light the course of the falling particles can be followed, and at the proper moment a sudden tilt empties the whole of the water and most of the foraminifera into the sieve, the sand and the heavier 'forams' being left in the jar. The purity of the material in the sieve, which is usually called 'floatings,' will depend upon the skill and judgment of the operator, and is largely a matter of practice. The residuum in the jar must be washed out into a basin for further treatment, and the operation repeated with more sand and water until the whole of the gathering has been treated. The time allowed for subsidence will vary with the fineness of the sand, so that in the case of the finest siftings nearly a minute may be required. The actual time can only be determined by watching the falling material in a strong light.

In the case of very fine sand, the tension of the surface film of water is so great that the sand grains float almost as readily as the foraminifera. This difficulty may be overcome by shaking up the contents of the jar, covering up the top with one's hand while so doing.

If it is desired to obtain a very pure gathering the jar should be stirred up and allowed to settle for a few minutes. The lighter species of foraminifera will be found floating and adhering to the sides of the jar at the surface of the water. They can be removed with a cigarette-paper or feather, and washed off into a separate small sieve. I call these ' double floats.'

The residuum, which had been set aside in a jar, may now be treated by the ' rocking' process for the separation of the remaining foraminifera. Taking the photographic developing dish (or a tin tray may be used as a substitute), enough of the residuum is placed in it to cover the bottom to a depth of about $\frac{1}{4}$ inch, and covered with about $\frac{3}{4}$ inch of water. If the dish is then rocked with a combined up and down and circular motion, the foraminifera will rise in suspension in the water, and by a little careful manipulation may be gathered in one corner of the dish. A sudden tilt will then empty them with the water into a sieve. The operation should be repeated with two or three lots of water, and the material left in the dish will then be found to consist almost entirely of sand. The material left in this second sieve, known as ' washings,' is not so pure as the 'floatings,' for it contains a large percentage of broken forms and shell fragments, coal-dust, and other débris. It may be further purified, if desired, by being dried and ' floated ' once or twice in the glass jar.

If the floatings thus obtained contain much animal or vegetable matter, as is sometimes the case, it is advisable to boil them in a solution of caustic potash. This will not damage the foraminifera so long as the boiling is not carried on too long, and it effectually removes the animal matter, which otherwise would encourage fungoid growths. This process must, however, be used with great caution if the gathering contains arenaceous foraminifera. All trace of the caustic potash must be removed by frequent washing.

The processes already described are intended for recent *sandy* gatherings. When the material is in the form of dredged mud, it is first necessary to get rid of the finest particles of this mud, for if the water is turbid it becomes very difficult to judge the right moment for separating the floating forams. The mud should be broken up into small lumps, about an inch cube, and slowly but thoroughly dried. It is then placed in a basin and covered with water, which rapidly breaks it up into a fine mud. Boiling water acts most quickly, and if the mud was very ' sticky ' when dredged the addition of washing soda will facilitate the cleaning process. Great care must, however, be taken not to expose your silk sieves to the soda solution for more than a few seconds, and to wash them thoroughly afterwards. Such specimens as may be observed floating on the surface of the water may be easily removed by means of cigarette-papers, which are placed on the surface of the water. The forams adhere to the papers, which are then carefully lifted off and dried, the specimens being afterwards brushed off into a tube. Many delicate forms, which would almost certainly be broken in the subsequent processes, may thus be obtained in a perfect state.

The mud remaining in the basin is then washed, a spoonful at a time, by placing it in a sieve of fine silk gauze, through which a gentle stream of water from the tap (or preferably from a ' rose ' attached to the tap by a short length of tubing) is kept running until all the fine particles have been removed. The muddy water should be allowed to settle in a bath, and the solid residuum scraped out and thrown away. The sandy residuum left in the sieve should be thoroughly dried, and is then ready for examination under the microscope, or, if desired, it may be further reduced in bulk and purified by the floating and rocking processes already described.

Foraminifera occur in marine fossil deposits of all geological ages, from the Cambrian to post-tertiary, but they are, as a rule, of sparing occurrence until we reach the cretaceous period. The harder chalks and limestones can only be studied by means of thin sections, but the softer chalks, shales, and clays may be broken up by drying the material in small pieces and washing it over a fine sieve in the manner just described. Floatings are seldom procurable from fossil deposits, owing to the weight of the specimens, which are generally more or less infiltrated with pyrites or other mineral matter.

Some chalks and shales which resist the disintegrating action of water after being dried may be broken up by the action of a crystallizing salt, which has been absorbed in a fluid state. Acetate of soda has the most rapid action, but very fair results may be obtained with common washing soda. The material, after being broken up into small pellets, is dropped into a boiling saturated solution of the salt, and kept at this temperature for a short time to allow of penetration. The salt is then allowed to cool, and in cooling crystallizes, the formation of the crystals breaking up the outer layer of the material. On being warmed, the soda dissolves again in its own water of crystallization, and the crystallization is repeated over and over again until the lumps are broken up. The resulting mud is then washed in the ordinary way.

The best foraminifera from the chalk are those obtained from the interior cavities of hollow flints. They are often in the most perfect state of preservation, and the chalk in these cavities being of a powdery nature, they are very easily cleaned.

The cleaned material should be sifted into varying degrees of fineness, and each grade kept separately in a tightly corked tube, noted with locality, date, and any details as to the species contained in it, which may be likely to be useful for future purposes of reference. If the material has been properly cleaned and dried it can be kept unaltered for an indefinite period, but if put away damp fungoid growth will quickly set in. This can be destroyed and the material sterilized by a prolonged soaking in spirit, the material being afterwards dried once more.

To examine the material under the microscope, a picking-out tray will be necessary. This is made by covering a slip of card with coarse black ribbed-silk, the ribs running longitudinally along the slip. A thin wooden ledge must be glued round three sides of the slip to prevent the forams rolling off when the stage of the microscope is inclined at an angle. The 'Stephenson' form of binocular is, of course, the most desirable for the work of selection, as the stage being always horizontal, the specimens cannot roll out of the field. The material is sprinkled over the slip, and the ridges of silk keep the forams from rolling about. The specimens required can then be easily selected by means of a fine sable brush, moistened by drawing it between the lips, and transferred to a prepared cell or slip.

The best fixative for mounting foraminifera is gum tragacanth, which is almost invisible when dry, being quite devoid of the objectionable glaze which characterizes gum arabic. It is also much less subject to variations of moisture than gum arabic, which alternately contracts and expands with changes of weather, and often fractures delicate forms. Powdered gum tragacanth should be used in the preparation of the mucilage. Put a small quantity of the powdered gum in a bottle with sufficient spirit of wine to just cover it. Add a small crystal of thymol or a few drops of clove oil, or oil of cassia, as an antiseptic, then fill the bottle with distilled water and set it by for some hours. The gum will form a thick mucilage, and may be used of varying thicknesses according to the size of the foraminifera. For most forms it should be of about the consistency of cream, and it may be used liberally in mounting, as it shrinks very much in drying.

The same gum diluted to a watery consistency can be used as a fixative for foraminifera mounted in balsam. If the slide is thoroughly dried before the balsam is added the gum becomes quite invisible.

For very large and heavy foraminifera, seccotine or some other liquid glue may be used with advantage, gum not being of sufficient strength to hold them safely.

Many fossil foraminifera and recent forms from some localities have the internal chambers filled with mineral infiltrations, either glauconite or pyrites. These internal casts reproduce more or less perfectly the shape of the sarcode body of the animal. They may be obtained by decalcifying the specimens with very dilute nitric acid, just faintly acid to the taste. To obtain perfect casts the process must be carried out very slowly, adding drop by drop to the watch-glass containing the specimen. When decalcification is complete the resulting cast should be carefully removed with a pipette, and deposited in a spot of gum on a slip. They will not stand transference with a brush without damage.

Artificial casts of the animal body may be taken in paraffin wax, by a method suggested by the writer to Mr. H. J. Quilter, a nd described by him in the *Journal of the Quekett Microscopical Club* for 1903.

Briefly, the method as improved by subsequent experience is as follows: (1) Boil the foraminifera in weak caustic potash to remove animal matter, wash thoroughly and dry. (2) Soak in chloroform for a few days, until all the air is expelled from chambers. (3) Drop straight from chloroform into melted paraffin wax of high melting-point, and keep the wax hot over a spirit lamp until all the chloroform has been expelled as bubbles. (4) Then remove the foraminifera with a forceps or brush to cover-glasses, and remove any excess of paraffin wax with blotting-paper. (5) After the wax cools, the foram is attached to the cover by a film of wax. Clean the top surface of specimen by rubbing it with a fragment of blotting-paper dipped in xylol. This operation should be done under the microscope, as it is only desired to remove the surface film of wax. (6) Drop the cover-glass into a beaker of strong nitric acid. The rapid effervescence tears the outer film of wax apart, and the internal cast floats free. It may then be taken up carefully on a brush and mounted dry.

The process is a most interesting one, and the results with the larger species are surprisingly beautiful. The smaller species are, however, very difficult to manage satisfactorily.

Sections of foraminifera may be made, and will be found very useful, and sometimes indispensable, for the study of their internal structure. If it is merely desired to lay the shells open in order to show the interior chambers as an opaque object, this may be done by fixing them on a slip or cover-glass and rubbing lightly on a fine hone or on a sheet of fine ground glass. The risk of fracture is greatly lessened if the chambers are first infiltrated with wax or with Canada balsam in the same manner as in the preparation of casts, the matrix being removed with xylol after the section has been made. If, however, it is desired to cut a transparent section, the chambers must be infiltrated with toughened balsam. After soaking in chloroform until all air is got rid of, the specimens should be placed in a watch-glass and covered with balsam in chloroform. This must be left for a day or two, then toughened in a cool oven. If the balsam is then melted the specimens can be removed to slips, where they will be fixed on cooling, and are then ready to be ground down.   When ground to the required depth the slide is again warmed, the specimen turned over, and the grinding repeated with extreme care until the section is reduced to the required thickness.   The operation is one of the most extreme delicacy, and no microscopist need expect to succeed without preliminary failures innumerable.   A single movement on the hone beyond the requisite point, generally suffices to wipe the specimen out of existence.

CHAPTER XXV

NOTES ON THE COLLECTION, EXAMINATION, AND MOUNTING OF MOSSES AND LIVERWORTS

By T. H. RUSSELL, F.L.S.,

Author of 'Mosses and Liverworts.'

Collection of Specimens.—The appliances for the collection of specimens are simple in the extreme. For many years I have been in the habit of putting the material gathered in the field into old envelopes that have been cut open at the narrow end instead of at the side. Not only do these form most convenient pockets for the purpose, but notes can be made on them at the time, of the date and locality when and where the plants were found, and of the names that suggest themselves on a first inspection (a useful aid to the cultivation of the power to recognize plants in the field). Rough memoranda, too, may be added afterwards of any special features of interest that present themselves on a closer examination of the contents, and that need elucidation. Squares of stout paper will be equally serviceable, and, indeed, some of these should always be included for the putting up of larger plants. They will also prove more convenient when specially wet material has to be stored, as the only objection to the use of the envelopes is that moisture is apt to loosen the gum which fastens the several portions together.

A magnifying-glass with a fairly large field, for making a preliminary acquaintance with the gatherings, and for exploring purposes, will also be required. I always carry a second glass of a higher magnifying power—*e.g.*, Browning's platyscopic lens with a magnification of ten diameters—which is often most helpful when minute details have to be ascertained. An old knife, with which to dislodge the plants and to free them from soil and grit; a lead-pencil for making notes of habitats, etc.; and a satchel, to serve as a receptacle for the envelopes and papers, both when empty and filled, complete the only real requirements of the moss-hunter.

On reaching home the envelopes that have been used should be placed in a warm room, in an upright position, and with the ends open as widely as possible so as to admit air, and the packets unfolded; this will allow the specimens inside to dry, a matter of no small moment if the risk of mildew is to be avoided. After standing thus for a few days they can be safely put away until a convenient opportunity occurs for examining the contents, when soaking in hot water will speedily restore the plants to their original freshness, though not, of course, to life. The small china saucers, made in different sizes, to be procured from any artists' colourman, are very convenient for this purpose. The ease with which mosses can in this way be revived for examination constitutes, to my mind, one of the chief attractions which this branch of botanical research offers to anyone in search of a hobby, for while the gathering of specimens forms a healthy outdoor occupation for all seasons of the year, and adds immensely to the pleasure and interest of a ramble in the country, their examination and mounting may be deferred for any length of time, and will provide the most pleasurable recreation by the fireside in the long winter evenings.

Preparation of Specimens.—To prepare specimens for examination or for mounting, some form of dissecting microscope is practically a necessity. For many years I used an ordinary magnifying-glass of low power, mounted in a light metal frame, at one end of which is a small collar, which slips over a screw fixed in an upright position in a small metal stand, and provided with a nut by means of which the lens can be adjusted to any required height, and this simple expedient is still often very serviceable.

Another inexpensive and useful instrument for the purpose is the ordinary watchmaker's glass, consisting of a lens set in a deep horn mount, by the help of which it can be retained in position in front of the eye; but care must be taken, especially when this is used in mounting, that the muscles are not unconsciously relaxed in the interest of the work, and the glass allowed to fall. As a rule, however, I now employ the more modern binocular dissecting microscope, which is also of the greatest assistance in mounting.

For dissecting purposes ordinary sewing needles set in cedar penholders are frequently very handy, and it is well to have one or two bent at an angle to the holder, as these will be invaluable in mounting, both for the purpose of altering the position of objects after the cover-glass has been put on and for removing stray bubbles of air. In order to bend a needle into almost any form, it is only necessary to heat it in a spirit-lamp to a red heat, and then allow it to cool; this will render it soft and pliable. After being bent to the required form, it can be rehardened by plunging it when red-hot into water.

While dissecting instruments of various kinds can be purchased, I know of none that better serve the purpose than such as may easily be constructed by the use of sail needles and glovers' needles (No. 4); the former for ordinary work and the latter for the more delicate operations. Sail needles can be obtained at an ironmonger's and glovers' needles from a draper. The needle, in addition to having a fine point, is ground with three flat faces, which give as many cutting edges. In the case of the sail needles these edges are somewhat blunt, and the needle should therefore be sharpened on a hone before being used. Cheap black-lead pencils make very good handles in which to mount the needles. After being cut into suitable lengths, they should be well soaked in hot water; this melts the glue that fastens the two portions of the wood together, and the lead can then be easily removed. The needle should be placed in the groove thus provided for it, and should be allowed to project some little way beyond the end of the wood. A fine pin or needle is then run through the eye, and is cut off with a pair of wire-nippers, leaving just enough to press into the other half of the holder, when put into place. Some amount of packing may be necessary, especially with the glovers' needles, to keep the needle rigid. The two parts of the holder are then fastened together with liquid glue or seccotine, each end being afterwards tied round with fine twine, as a still further precaution. It is well to scrape off all traces of the glue from the outside of the holder, as otherwise it is apt to stick unpleasantly to the lips if held there for a minute in the hurry of mounting. Two pairs of forceps (one with curved ends), a pair of small scissors, a small camel's-hair brush, and one or two small lancets, will practically complete the implements required for all ordinary dissecting, to which must be added, for the purpose of microscopical examination and subsequent mounting, a stock of the usual glass slips (3 inches by 1 inch), and cover-glasses of two or three different sizes. Though the round cover-glass is generally to be recommended, both on the ground of appearance and of ease in sealing, yet with many of the larger mosses the square form will prove more serviceable, as it gives more mounting surface, while for dealing with plants of any considerable size specially large pieces of cover-glass may have to be procured.

Mounting.—Owing to their small size and the facility with which their original freshness can be revived, as already noticed, mosses can be far more satisfactorily preserved than is possible with ordinary flowering plants. The greater number may be readily mounted on the ordinary glass slips, and in this form they not only occupy a comparatively small storage space, but remain, for all practical purposes, as fresh as when they were gathered. I have specimens in my collection now that were put up twenty years and more ago, and which have altered little in appearance in the meantime.

I have tried several materials and compounds for mounting purposes, but unhesitatingly give the palm to glycerine jelly, both on account of the ease with which it may be manipulated and by reason of its admirable preservative powers. I shall, therefore, mainly confine myself to a description of the method of procedure when this medium is used. I have for many years made my own jelly according to the following recipe, which is a slight variation on that given in Carpenter's work on "The Microscope." Take 2 ounces (by weight) of the best gelatine, 6 ounces of water (also by weight), and 6 of glycerine. Soak the gelatine in the water until it swells (this takes about forty minutes); then place the vessel containing the gelatine and water (a jam-pot is very serviceable; it should be provided with a cover of some kind) in a saucepan of water, and boil over a slow fire until the gelatine melts. When the gelatine is cool, but still liquid, add the white of one egg, and mix well. Boil the gelatine as before, until it becomes thick with the coagulated albumin—this takes about twenty minutes ; add the glycerine and 25 or 30 drops of carbolic acid and mix well ; strain through filter-paper before the fire, and a clear pale yellow jelly should be the result.

The specimen to be mounted must first be cleansed from all earth and grit in water, the spores gently expelled from any capsules, and all air-bubbles removed by means of the dissecting needles, and too much care cannot be given to this somewhat tiresome process, as on its due performance the success of the mount, to a very large extent, depends. But our moss is not yet ready to be put into the jelly, for if this were straightway to be done, the effect would inevitably be that the leaves would curl up beyond recognition. A simple plan for guarding against this is to boil the plant for a few seconds over a spirit-lamp, in a teaspoonful of water, in which three or four drops of glycerine have been mixed ; but unless time is of moment, it is better to leave it to soak for twenty-four hours in a mixture composed of water, $1\frac{1}{2}$ ounces (fluid) ; rectified spirit, $1\frac{1}{2}$ ounces ; and glycerine, 5 drachms. The small china pans in which moist water colours are sold are very useful for this purpose. Not unfrequently I subject it to both treatments, as this tends more thoroughly to remove all air. When taken from this preparatory bath care must be taken to remove the fluid adhering to the plant as far as possible ; this may be done by placing it on a glass slip, and tilting this so as to allow the superfluous liquid to drain off, with possibly a judicious application of blotting-paper, though this must be used with caution, as otherwise fresh air is apt to be admitted.

A hot-water bath is essential for mounting with glycerine jelly. A simple and inexpensive one can be made by means of a small glass tumbler, provided with a closely fitting tin cap or lid, having a piece cut out of the margin, leaving just room enough to admit the neck of a small glass bottle containing the jelly. The bottle can thus hang in the hot water in the tumbler, when the lid is in place, by means of its projecting lip, which rests on the top of the tin cover, and in this way the jelly is kept melted, and is, moreover, close at hand for use. The advantage of having a *small* bottle for this purpose (which is, of course, replenished from time to time from the larger stock bottle) is that the necessity is obviated of continually remelting the same jelly, which would cause undue evaporation of some of its component parts. When my mounting is likely to take long I wrap a piece of flannel round the tumbler, in order to retain the heat in the water as long as possible. The glass slip on which the mount is to be made, as also the cover-glass, must be first carefully cleaned. A good plan is to rub them over, between the finger and thumb, with acetic acid, in order to remove all traces of grease, and then to wash them in warm water, afterwards drying with a soft cambric handkerchief, a final polish being given with a wash-leather. The glass slip is now placed upon the flat tin cover of the hot-water bath, a small pool of the liquid jelly is put upon it by means of a pipette, and the specimen, after being freed from the preparatory fluid, is gently lowered into the jelly. While the latter is kept liquid by the heat of the water-bath, all air-bubbles must be carefully removed with the dissecting needle, and here the binocular dissecting microscope will be found most helpful.

It is impossible to exaggerate the importance of this extraction of air, for nothing detracts more from the appearance of a mount when viewed under the microscope than the presence of these disfiguring silvery globes, lurking among the delicate leaves, or perhaps in the teeth of the peristome ; and my own rule always is that, rather than allow a serious blemish of this kind, the slide must be sacrificed or the mount be recommenced. I have found it a great help in many cases, especially when an object likely to retain air or an undue amount of the preparatory fluid is in hand—as, for instance, a large empty capsule or a plant with the leaves closely covering the stem—to put it into a little jelly on a spare glass slip, and then to extract the air as far as possible before transferring it to the slip on which it is to be mounted. The whole plant thus becomes more or less saturated with the melted jelly, and the air-bubbles cannot find their way back to the mount, as they are apt to do if the whole process is carried out on the one slide. A second hot-water bath is then a practical necessity, in order that the jelly thus left in contact with the specimen may not solidify before the latter is transferred to the glass slip. When everything has been prepared, and the specimen is in place, immersed in plenty of the liquid jelly, the cover-glass is taken up with the forceps, and gently lowered on the jelly, beginning from the left-hand side, driving the jelly (and too often, alas! the specimen also) before it, as it is allowed gradually to fall into place. This is an operation of no little delicacy, as if great care is not taken a large bubble of air will make its way in at the last moment. If, as frequently happens, the putting on of the cover-glass has caused a displacement of the object, this must be rectified before the jelly is allowed to set, and here the bent dissecting needles will be of great service, as a considerable amount of rearrangement can be effected with one of them, and stray air-bubbles may also be removed without disturbing the cover-glass.

Should it happen that not quite enough jelly has been used to fill the whole of the space under the cover-glass, a small additional quantity must be introduced by means of the pipette, care being taken to avoid the formation of air-bubbles. The slide is now taken from the hot-water bath and is allowed to cool, and in a few minutes the jelly will have so far solidified that it can be examined under the microscope, when, should any serious defect be disclosed, the jelly must be remelted, and the short-coming be rectified. The final process consists in removing the superfluous jelly from around the cover-glass with a knife, cleaning the slide from all trace of the jelly (a handkerchief moistened at the lips is the most efficient method), and sealing the cover-glass round the margin with some kind of varnish.

I may add that I usually mount two cells on each slide; in the larger of them I place a small portion of the moss, together with a few capsules, if possible in various stages of growth, and two or three perichætial leaves, while the other cell contains some leaves dissected from the plant (where of importance from both stem and branch), and a few pieces cut from the mouth of the capsule, to show the peristome.

It will be very evident that the subject of the removal of air has frequently cropped up in what I have written; this is accounted for by the fact that it betokens the chief difficulty to be encountered in mounting when glycerine jelly is the selected medium. There is, unfortunately, no royal road to success in this particular, and the main thing to rely on is a patient use of the dissecting needles. One or two hints as to matters of detail may, however, be given. Boiling the specimen, more especially if it be old, is often helpful; and soaking it in water that has been allowed to *boil* for ten minutes is sometimes recommended. In this connection, too, I may mention that it is important to keep the pipette, used for abstracting the melted jelly from the bottle, quite clean, as, if earthy matter is allowed to get encrusted on it, small bubbles of gas are apt to be formed, and these easily get transferred to the jelly itself. I have also found that attention to another seemingly trivial point is of no little moment, and that is, when dipping the pipette into the melted jelly always keep a finger on the open end until the point has reached the bottom of the bottle. As heat naturally causes any bubbles in the jelly to rise to the surface, this minimizes the risk of their entering the pipette, for the lower strata of jelly are, so to speak, tapped.

Glycerine Jelly.—As far as my acquaintance with glycerine jelly goes—and it is one of a good many years' standing—the great objection to its use arises from the fact that after a specimen has been mounted in it, and has stood possibly for years, the jelly may develop an unpleasant tendency to liquefy, as will be evidenced by the presence of small beads of glycerine round the edge of the cover-glass.

While not being able to suggest any unfailing remedy for evils such as this, I have noticed that the observance of a few simple rules will considerably minimize the risk. These are—

1. Care should be taken to remove the preparatory fluid from the surface of the object as far as possible before mounting, without, of course, running too much risk of admitting air.

2. No pressure should be applied to the cover-glass in order to keep it in position, its own weight being alone sufficient for the purpose.

3. A sufficient quantity of the jelly should be used to allow of a small portion extending on every side beyond the edge of the cover-glass, in order to provide against subsequent shrinkage in the gelatine.

4. The slide should stand for at least two or three months before the additional jelly is removed and the cell sealed.

5. The slides should be kept in a fairly equable temperature, and should not be exposed to draught.

I generally add a very small amount either of carbolic acid, or of a 5 per cent. solution of bichloride of mercury (corrosive sublimate) to the liquid jelly on the glass slip, before the specimen is put into it, as this lessens the chance of any fungoid growth subsequently developing in the cell. It will suffice if the tip of a thin glass rod that has been drawn to a point is just dipped into the liquid, the small amount clinging to it being transferred to the jelly, and quickly mixed with it. In the case of the bichloride of mercury it is especially important not to introduce too much, or a slight precipitate of calomel will result, giving a cloudy appearance to the jelly.

Formalin.—A 3 per cent. solution of formalin is also a most serviceable mounting medium, more particularly where some of the more delicate plants are under treatment, though it is, of course, open to the objections that attach to the use of all liquid media. If only leaves are being mounted—and it is for such a purpose that the solution is most suitable—a cell of some spirit varnish must first be made by means of a turn-table. When this is dry the top of the cell is ringed round with marine glue dissolved in benzol (as to which, more hereafter), and sufficient of the solution is introduced into the cell from a pipette to allow of its standing well above the walls of the cell. The object is now carefully introduced, the cover-glass is taken between the finger and thumb, and, after being brought as near to the solution as possible, is allowed to fall gently into place. Should air have unluckily made its way in, the cover-glass must be quickly raised, and a little more of the solution be introduced. The cover-glass is now gently pressed down on to the top of the cell, and, after all superfluous moisture has been removed with a handkerchief, is ringed round with the marine glue solution, and afterwards with spirit varnish ; being finished off, if thought desirable, with a coat of asphalt varnish. If a larger object is to be mounted a deeper cell must, of course, be used.

Varnishes.—I have tried a good many sealing materials, and on the whole much prefer picture copal varnish (to be obtained from any artists' colourman) thinned with benzol. It does not dry too hard, and in consequence is not liable to crack ; while, should any portion of the object happen to be located near to the edge of the cover-glass, it can nevertheless be seen through the practically colourless varnish. Where the object is of any size, such as a piece of one of the larger plants, it becomes advisable to use a sealing medium that will adhere more tenaciously to the sides of the cell, and here nothing serves the purpose better than marine glue dissolved in benzol. I should advise anyone who intends to employ it to get the preparation ready-made rather than attempt its manufacture, for if there is one medium that is more exasperatingly adhesive and sticky than another it is marine glue. It is, however, a very safe material to use, and is, as far as my experience goes, quite free from any tendency to ' run in,' which I have always found to be the shortcoming of gold size. The cell may be finished later with asphalt varnish.

Although I have throughout referred to mosses alone, yet the methods of which I have spoken are equally applicable to the mounting of liverworts. The only additional point to be mentioned with regard to the latter is this : Owing to the fact that the contents of the leaf-cells are specially dense, it is often advisable, in order to render the cell-structure more distinct, to treat some leaves with a strong alkali. The best way is to place a piece of the plant in a few drops of a 7 per cent. solution of liquor potassæ on a glass slip, to cover this with a cover-glass, and then to boil over a spirit-lamp. The plant will need to be well cleansed by boiling in water before the leaves are mounted and a few of them may then be conveniently included in one of the cells.

CHAPTER XXVI

THE MICROSCOPE AND NATURE STUDY

By WILFRED MARK WEBB, F.L.S.,

*Editor of 'Knowledge' and Honorary General Secretary of the
Selborne Society.*

YEAR by year the interest which is taken in the world around us, in the unspoiled works of Nature, continues to increase. It is now also generally recognized that to train the powers of observation is one of the most important necessities in general education, and that it is far better for everyone to teach themselves naturally through the interest aroused by the subjects considered, than to learn nothing but second-hand facts from others.

To the majority of people, whether they are children or not, living things prove most attractive. The general appreciation of them begins most easily and properly out of doors, and may continue as a lifelong pursuit. The detailed investigation of some part of natural history forms an interesting hobby as well as a healthy form of exercise, recreation, and relaxation, particularly for those who are not forced to spend all their spare time upon games, or to whom the more violent forms of athletics and the usual kinds of sport do not appeal.

Whatever line of study is taken up, it will be very soon found that if any real progress is to be made—if anything new is to be found out with regard to the structure of the various creatures apart from the obvious—some aid to the vision must be sought, some means of learning details which cannot be seen with the naked eye. Here it is, then, that the microscope comes into play, and it is not too much to claim that besides being the source of additional interest, the instrument is a great educator

—that is to say, it trains, without appreciable effort, the hand to be skilful, the eye to appreciate, and the brain to elucidate. Moreover, without for one moment suggesting that observations in the open air should not be considered the most essential part of Nature study, we must agree with a recent writer in *The Country Home*, that 'there will be times when the most enthusiastic Nature student cannot be out of doors—long dark winter evenings and wet days even in summer, when indoor work must take the place of outdoor. It is then that work with the microscope will prove such a fascinating hobby, supplementing, as it does, the observations made with the naked eye, and leading us into regions where it is impossible to travel without it.'

The lowest forms of plant-life are unicellular, and often extremely small. The microscope reveals to us that they have powers of locomotion ; it shows us also, for instance, that the green colouring on trees and fences which shows after rain is made up of myriads of minute plants, taking in gases from the air and earth-salts from the surfaces on which they live, and making their food in the same way as the cabbage or the oak-tree. The whole science of bacteriology and the discoveries of the minute fungi which cause disease and putrefaction, which give the taste to butter and the flavour to cheese, entirely depends upon the high powers of the microscope, and though investigations in this direction are not for the young beginner, still, they lie before him when he has mastered the details of his instrument.

The story of the interesting, though for a long time hidden, methods of reproduction among the mosses and ferns have been revealed by the microscope. The eggs and motile fertilizing bodies in the so-called flowering heads of the moss have been discovered, and the determination of the species by the systematist depends to a great extent upon the microscopic details of the capsules which grow from the egg, and are really another generation, producing spores without fertilization and getting its nourishment from the original moss-plant. In the ferns and their allies, however, we find the sexual organs are borne by a tiny plant like a minute liverwort, which springs from the spores on the fern frond when they fall to the ground, and this little plant nourishes the egg as it develops into a new fern, which, unlike the green moss-plant, is the sexual generation.

Before going on to speak of the use of the microscope in the various branches of natural history, we may point out that it can with advantage be used from time to time to lend an additional interest to ordinary Nature study.

The youngster who sees the pollen of various kinds of flowers as a powder may well be introduced to the variety of shapes and sculpturing which the grains present. Many of the hairs which clothe and protect the commonest plants are fascinating when their details can be seen.

An unfortunate occurrence such as the stinging of a youthful naturalist by wasp or bee may well lead to the examination of the sting of the insect, and possibly the hairs of the nettle, while the delicacy of natural objects compared with those made by man may well be brought home by examining the point of a fine needle with a microscope, and seeing how far more clumsy it is than either of the other two structures to which allusion has been made.

We need not deal further with this side of the question, for those who look for suggestions as to microscopic work, can glean them from the paragraphs in which more systematic work is discussed. It may be pointed out here, however, that the microscope may be used by the young student so soon as the informal stage of Nature study is passed. The writer can say from personal experience that the interest which can be aroused is very great, while excellent work with the microscope has been done, for instance, by gardening lads who for years have used no instruments of greater precision than spades, and rakes, and hoes.

We need not dwell any longer upon the botanical side, except to say that the whole structure of plant bodies lies before the student. There are all the interesting details to be worked out in connection with the formation and storage of starch grains, which vary in different plants; while one may examine the delicate hairs on the roots which take up water and food materials, or the thread-like fungi which sometimes enter into partnership with the roots and do the work of root-hairs, as in the heaths and rhododendrons, not to mention the bacteria-like organisms which produce nodules on the roots of plants belonging to the pea family, and supply their willing or unwilling hosts with nitrogen which they are able to get from the atmosphere.

Then there is the structure of the fibres which prevent stems from breaking, and of the tubes which carry water with all the various kinds of thickenings which strengthen their walls and prevent them from being crushed in as the plant grows and the pressure within the bark increases.   There is the bark itself, made up of many empty brick-shaped cells, and the places known to botanists as lenticels, where the bricks are, as it were, heaped together, instead of making a solid wall, so that air can penetrate to the living tissues below.   With the microscope we see how the annual rings come to be made in a woody stem.   We can learn the structure of a bud and the growing tip, and really come to know how a plant is built up.

If the living plants should pall, we can cut thin slices of fossils and trace the affinities between plants of bygone times and those of the present day.   In fact, there is no end to the beauty and the interest that is revealed by the microscope when it is brought to bear on plant structures.

If the material offered by the animal world is not more varied, it is, if possible, even more attractive than that which is to be looked for among plants.   To be sure, the botanists have the beautiful flinty shells of the diatoms to study, but among the unicellular animals there is a wealth of forms which are provided with calcareous shells of many shapes, or which build them up with the marvellous discrimination which may exist even in a microscopic speck of protoplasm, from sand grains, or the flinty needles of sponges.   It is these shells of foraminifera which, to a large extent, form the ooze which is taken from the very greatest depths of the ocean.

Other slightly higher forms have internal silicious skeletons, and may be caught living in fine nets or their skeletons obtained from deposits such as the Barbados earths, which are largely composed of them.

There are many of the creatures, such as the bell animalcule, the slipper animalcule, and any number of other Infusoria whose conformation appeals to the eye and whose life-histories are fraught with interest.   In the sponges our microscope tells us that amongst the supple horny fibres or the delicate needles and geometrically formed spicules which make up the skeleton there are small chambers in which lie the working cells that differ in practically no respect from Infusoria. A step, however, takes us to the polyps, and a favourite subject for study is the little fresh-water form, with its waving arms, covered with sting-ing thread-cells that aid it to obtain its prey. Its body, which is all stomach as it were, is lined with cells resembling the proteus animalcules of the ponds, but, unlike them, unendowed with the power of individual locomotion.

For charm of shape and delicacy of construction commend us to the skeletons of the polyps, which live in colonies, to the fixed growths on rocks and sea-shells which at first sight look like delicate seaweed, but which on examination are seen to bear innumerable little cups in which the polyps are seated. Almost microscopic, too, are the small jelly-fish which bud off from these colonies, and by an alternation of generations reproduce not them-selves, but colonies like those from which they sprang. Passing over the hedgehog-skinned creatures covered with little nipper-like projections or spines, whose internal structure is of suffi-cient beauty to repay the trouble of grinding sections, and leaving on one side the wheel animalcules, pretty Polyzoa, and the host of creatures known as worms, which offer many problems to the biologist, we come to the molluscs.

Their shells are usually large enough to see with the naked eye, but there is a great fascination about the structure of their calcareous coverings. Simple though this may be in the bivalves, it is intricate and puzzling enough in the univalves to satisfy the demands of those who wish to exercise their brains and ingenuity. Then the examination of the tongue-like organs of the slugs, and whelks, and limpets is an aid to the classification of these forms. These structures themselves, covered as they are with minute rasp-like teeth, are so very varied and beautiful that the pleasure of examining them, apart from their scientific investigation, can be well understood. The true snails, again, are often provided with minute calcareous spicules of character-istic shape in the various species, called the 'love-darts,' which are useful for classification purposes, and form beautiful objects for display.

It is when we come to the insects, however, that we meet with

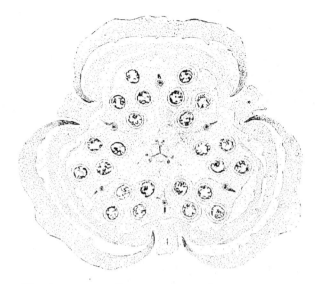

FIG. 87.—SECTION OF FLOWER-BUD OF LILY, SHOWING OVARY
ANTHERS PETALS, POLLEN GRAINS, ETC.

FIG. 88.—LOPHOPUS CRYSTALLINUS. FIG. 89.—STEPHANOCEROS EICHHORNII.
By permission of Messrs. Flatters, Millborne, and McKechnie, Ltd.

[To face p. 296.

even more unlimited material. Some of them, like the tiny fairy-flies,—of which, according to Mr. Enock, five can walk abreast through the hole made in a piece of paper with a pin-point—are so small that they have to be examined under the microscope in their entirety, while the parts of other insects show a wealth of detail, and illustrate in a marvellous way the changes which Nature can ring on a single plan. Take, for instance, the mouth-organs of the cockroach. At first sight they little resemble those of the bee, which sucks rather than bites, and appear to have no connection with the proboscis of the butterfly, intended merely for drinking up honey. On careful examination, however, it can easily be seen how the mouth-organs of the two latter have been modified from the first, and a similar comparison may be made with the stylets of the flea, the piercing organs of the bug, and the lancets of the gnat.

It is only when we examine the wonderful proboscis of the fly that a real difficulty arises. We may mention also the beautiful scales which give the colours to butterflies, and which resemble those found on the more lowly wingless insects known as spring-tails and bristle-tails, which have never known what it is to fly, and have only survived because when their relatives took to an aerial life, they were thrown out of competition with them.

There is, indeed, no end to the work which can be done on insects—their breathing apparatus consisting, as it does, of a series of tubes, from which the air is laid on, as it were, all over their bodies ; their beautiful antennæ, and the joints of the legs by which beetles, for instance, are recognized, offer fields for inquiry and objects of interest of which the student will never tire. The dexterous dissector will find full scope for his powers when unravelling the organs of insects, and finding out how these are equally well adapted to the requirements in the smaller forms of life which possess them, as are those of the higher animals, whose general anatomy it needs no microscope to elucidate.

In the vertebrates, as in the case of all living things, the minute structure must be learnt from the microscope, and, though the sections are more troublesome to obtain than in the case of vegetable tissues, there are a host of things that can be examined and worked out. Among them we may mention the scales of fishes, the hairs of animals, the feathers of birds. Everyone can see for himself how the feather is built up; and although all the larger feathers are made on the same plan, there are differences in detail. The various birds, indeed, might occupy the attention of a lifetime.

It is not needful to dwell any further on the question of the use of the microscope to the student of Nature, or what lies before those who decide to take advantage of it. In conclusion, one may say that it should be part of everyone's education to learn how to use a microscope, and to have some knowledge of the minute details of the living world.

CHAPTER XXVII

THE MICROSCOPY OF FOODS

By CUTHBERT ANDREWS, F.R.M.S.

THE application of the microscope to the examination of substances in common use, as of foodstuffs, may be considered in two more or less distinct ways. In the first the instrument is used to reveal the beauties of structure of the various tissues, and in this direction alone many interesting facts are brought to light. But the microscope has a more utilitarian mission—namely, the detection of adulteration and abnormal conditions. This function is by far the more important, and it is mainly in this direction that we propose to consider food-microscopy in these pages.

It is only of recent years that the microscope has attained its present high position in analytical work, but it is now fully understood that with a modest equipment most important facts may be readily established, particularly where only a very small quantity of the substance in question is available. In such cases the microscope becomes even more reliable than an ordinary chemical examination, as for the latter we almost always need a fair quantity of material on which to work ; whereas with the microscope the minutest portion may be subjected to the action of reagents, the result being equally reliable with tests made on larger quantities.

But in employing the instrument in this detective capacity it must be always remembered that, valuable as a microscopic examination may be, it is desirable, where possible, that it be confirmed and amplified by ordinary chemical methods.

Microscopic investigation has one great advantage. In the hands of an experienced observer it is quick, and often serves to put one on the right track, where a chemical analysis may take some time to afford similar data. Further, the microscope, like the camera, ' does not lie '—much, and, given reasonable care, reliance may be placed on its evidence. For instance, if a sample of intimately mixed starch-grains were submitted to a chemist, we think he would not be disposed to offer any opinion as to their origin on the information afforded by a chemical examination. But a microscopist of experience in this class of work could, with a fair degree of certainty, give the proportions of the various starch-grains composing the sample.

On the other hand, a microscopical examination of a film of milk would be no guide to the fat-contents of a specimen; whereas by exceedingly simple chemical means a figure accurate enough for all practical purposes may be readily obtained.

We must therefore, on the whole, regard the microscope as an invaluable assistant, to go hand in hand with other and older methods. The chief use of the instrument, as above stated, is to aid in the detection of foreign substances in foodstuffs, and where the tissues present marked characteristics the visual examination is highly important. Further, the microscope is paramount when an examination for Entozoa or other parasites is to be undertaken.

Within the narrow limits of these pages we cannot pretend to give an exhaustive account either of characteristic tissues, their adulterants, or of microscopic methods; but we hope that the brief outline furnished may be of assistance to the working microscopist, for whose further guidance we append a list of standard works which should be consulted, and to the authors of several of which we have to acknowledge our indebtedness.

A word as to the choice of a microscope for this particular class of work may be helpful. In practice we find the ' H ' Edinburgh Student's, by Watson, an ideal instrument. It has every convenience for research work, and is also admirably suited for photo-micrography, which latter is often of great assistance in systematic work. With a couple of eyepieces and the 1-inch and $\frac{1}{8}$-inch objectives much may be done ; while the addition of a 2-inch and $\frac{1}{2}$-inch, with, later, a $\frac{1}{12}$-inch oil immersion, makes the outfit complete and efficient. All the photographs reproduced here were taken by the writer with this instrument.

PLATE III.

Fig. 90.—Starch of Wheat.

Fig. 91.—Starch of Rye.

Fig. 92.—Starch of Barley.

Fig. 93.—Film of Butter.

Fig. 94.—Penicillium Glaucum.

Fig. 95.—Cheese Mites.

As a lower-priced alternative to this stand we favour the 'Fram,' by the same makers. Having no mechanical stage, this is less convenient than the 'H'; but, after all, such a stage is largely a matter of habit, and readers will find the 'Fram' excellent and reliable in every way.

We may now proceed to a consideration in detail of some of the more important foods.

Starches.—Starch is the carbohydrate content of vegetable cells, and is found in the majority of plants. It occurs in various parts of the growth. To the human body starch is of considerable value as an energizing food, and in one guise or another it forms a large percentage of the nutriment we take. Immediately on entering the body it undergoes a change, being converted into sugar by the action of the saliva, and is thus prepared for its ultimate digestion and assimilation. The study of starches is therefore a very important part of food-microscopy, and considerable training is necessary in order to be able to state with certainty the origin of an unnamed starch.

Continued research has shown that starches possess some very decided characteristics. Thus, although different varieties vary considerably in size and apparent structure, yet the grains of a given starch are remarkably constant in their appearance, or, alternatively, show variation within certain well-defined limits.

The simplest method of examining starch under the microscope is to take the smallest possible quantity and place on it a drop of distilled water, then applying a thin cover-glass. Such films are, of course, only temporary; good permanent mounts of starch-grains are very difficult to prepare.

Viewed by transmitted light, a film of starch presents a field covered with round, oval, or slightly irregular bodies, devoid of colour, and highly refractile. There are usually to be seen both large and small grains, excepting in the case of certain varieties, where the individual cells are remarkably uniform in size. Some starches exhibit concentric striations or markings, which may occur either with or without a 'spot' or hilum, the latter being either central or excentric. Alternatively, a grain may show no visible hilum or striation. These markings are obviously one method of distinguishing certain starches from some others.

An examination should also be made by polarized light, using a selenite. By this means the markings are brought more strongly into view, and other phenomena, such as the ' cross ' on various grains, aid in the identification.

The most important starch from a dietetic standpoint is that of wheat (Fig. 90). The grains are large, round or oval, and do not exhibit the markings above mentioned. There are also present numbers of small grains scattered through the field, but sizes intermediate between the two are rare. Wheat-starch may thus be distinguished from rye (Fig. 91), which is similar, but which commonly has many grains showing deep-rayed clefts and ragged, broken edges ; and from barley (Fig. 92), which it again resembles, but the grains of which are rather smaller and not so uniformly round.

Bread consists mainly of starch, and if a small fragment of a white loaf be taken, and a drop of the iodine solution (p. 317) applied, a blue-black coloration will take place. This reaction is peculiar to all starches, and is a sure means of demonstrating their presence. It is also of value in examining ' pre-digested ' breads and cereal foods, as the proportion of unchanged starch may be estimated with a little experiment. Obviously, the larger the amount of starch, the deeper and more decided the colouring.

In most flours small fragments of bran are to be found, and these particles will be readily noticed under a moderate power.

Flour (and, of course, bread) usually contains also some proportion of aleurone, the proteid substance of plants. This will be differentiated by the iodine test above mentioned ; for, while the starch assumes a bluish tinge under the microscope, the aleurone grains take on a yellowish-brown colouring. A confirmatory method of identifying this and other proteid substances is by the application of Millon's fluid (see p. 317), with which the proteids give a ' brick-red ' reaction.

A further test of bread should be for acidity. This is indicated by a solution of litmus, a drop or two of which should be added to the specimen. If the latter is acid, the liquid turns red ; if alkaline, blue; while a neutral sample leaves the fluid unchanged. Good bread should show no acidity, even after keeping for several days.

In the case of ' brown ' or wholemeal breads, a lesser proportion of starch will be found, with a corresponding increase in the amount of proteid and bran.

The adulteration of white flours would probably take the form of an addition of starches other than that of the grain represented. Some, such as potato, would be at once conspicuous, by reason of the marked contrast in size. Other mixtures would be much more difficult to detect; and in the case of, say, a wheat-flour adulterated with barley or rye, it might be necessary to measure a large number of grains, and to compare the average of these with that of a specimen of known purity.

Alum has been used as a preservative in bread, to which it also imparts a whiter appearance, but it is probably seldom employed now.

Butter. — Under ordinary transmitted light, pure butter normally presents a nearly homogeneous, colourless background (fat), with the water-globules standing out prominently in the field (Fig. 93). These globules should show but little variation in size.

Viewed by polarized light, the field should appear uniformly dark, and a better examination may be made if all top light is screened off from the surface of the slide. The presence of salt or salicylic acid (occurring as preservatives) may occasion bright appearances, but these will be identified on examination by direct light, by reason of their refractile nature.

In the case of margarine, or foreign fats mixed with genuine butter, a more granular field is seen, and large water-globules are often present. Further, the polariscope will probably show many bright crystalline points; these latter are also observable in samples of rancid butter, or butter which has been reworked or ' renovated.' The number and size of the water-globules also vary with different samples, and may be an indication of the origin and method of manufacture.

Butter does not commonly contain substances other than fat and water, and, if present, such additions would be detected without much difficulty. Colouring matter, and gross adulterants such as starch, would stand little chance of being passed over even by a cursory examination.

To view butter conveniently, a very small fragment should be placed on a slip, and covered with a drop of pure olive oil. A thin cover should be very carefully pressed down on this, and a workable film, capable of being used under a ⅙-inch, will be thus formed.

The very greatest care must be exercised in dealing with butter samples. The art of 'blending' and 'faking' has reached a high pitch, and very scientific methods are now employed by some manufacturers. The size and arrangement of the water-globules; the presence of, or freedom from, salt; the degree of freshness of the sample, and many other conditions, will affect the microscopical appearance; and it is possible for a butter to contain a very high percentage of fat other than milk-fat, and for this to be still undetected by the microscope.

Cheese. — Cheese is not very distinctive when examined microscopically, the most important substance detectable being starch, which may occur as an adulterant. Very thin slices of the cheese should be cut, and the fat dissolved out with ether. The starch and proteid may then be demonstrated by the iodine and Millon's tests respectively.

In old cheeses the various moulds and parasites may be identified—*Aspergillus glaucus* (the usual 'blue mould'), *Penicillium glaucum* (Fig. 94), etc., and the common cheese-mite, *Acarus domesticus* (Fig. 95). The rinds of some cheeses are interesting, but, as this part is seldom eaten, they are not very important.

Milk.—It has been stated already in these pages that the microscopic examination of milk is of small service in determining whether the sample be 'good' or ' poor '—that is to say, whether the percentage of fat is up to the normal standard. Certainly, a grossly diluted sample might present a field in which the fat-globules were noticeably few, but in such a case a simple physical inspection would reveal the lack of quality equally well. In any case, such a test would be quite insufficient in itself, and the estimation of 'fatty solids' must therefore be referred to chemical analysis.

But when we come to inspect the nature of any sediment

which may be thrown down in milk, we find the microscope of unique service. Needless to say, if milk is properly dealt with in the dairy and during distribution, the deposit should be practically nil; but as the desired care is frequently not exercised by the vendor, it is always advisable to make an examination, which may be on the following lines. The milk should be either centrifuged or allowed to settle in a conical glass, the drop taken from the point of the vessel, and covered with a thin cover, as suggested for the examination of water (p. 307). A film suitable for viewing with a high power may be easily obtained.

The microscopic appearance of normal milk is simply that of an almost colourless liquid, in which may be seen floating the fat-globules. The number of the latter will be under the normal in a film taken from a sample which has been allowed to settle; but it is the foreign substances in the field which are of most interest to the microscopist. Amongst the more important of these will be particles of yellowish-brown vegetable tissue, for the presence of such débris is a pretty sure indication of manurial pollution. This means carelessness on the part of farm-workers and others handling the milk, for which there is no excuse. A similar reason will usually explain the occurrence of grit, hairs, and fibres, and sometimes fragments of hay and pollen-grains. All these things indicate the possibility of the germs of disease being introduced into the milk; and as, with reasonable care, it is possible to avoid running this risk, a supply of milk which presents these danger-signals should be vetoed as human food.

Of matters other than those of vegetable origin the most significant are corpuscles of blood and pus. These are found in the milk of cows suffering from tubercle or from an inflammatory condition of the udder or ducts. The blood-discs of the cow closely resemble those of man, being only slightly less in size. The pus cells, if derived from an inflamed udder, or more remote locality, may be infiltrated with milk-fat, and in this state closely resemble colostrum corpuscles.

These last-mentioned bodies are found in the milk of cows which have recently calved. They are large, round masses, of a granular appearance, somewhat resembling a mulberry, and are believed to be really white blood-corpuscles (leucocytes) infiltrated with milk-fat. They are coloured very slowly with aniline red.

As the milk from cows which have newly calved should not be taken for human consumption, it is obvious that the presence of colostrum bodies again means carelessness or disregard on the part of the farmer.

The detection of preservatives is a fairly simple matter, but as this is best done with a larger sample of the milk, it scarcely comes within the range of our inquiries. Many of the customary tests could, however, doubtless be applied to microscopical quantities if necessary.

The bacteriological examination of milk is a very important study, and one which is daily becoming more imperative. No doubt exists as to the danger of milk as a carrier of disease; and this is readily understood when we realize how well suited the product is to the rapid multiplication of micro-organisms. Films may be prepared from the sediment, and stained for tubercle bacillus as directed by Mr. Cole, and, if necessary, cultivations should be made on a suitable medium. The germs of typhoid, scarlet fever, cholera, and diphtheria may also be present, and a textbook on bacteriology should be consulted for the best way of searching for these.

Fig. 96 illustrates the fat-globules in a film of pure milk.

Water.—Chemically pure water is, outside a laboratory, an impossibility. Moreover, it is highly probable that such a water would be of little or no use as a food; for it is not always remembered that water is needed in far greater quantity than any other foodstuff, although this is easily understood when we recall the fact that the body itself consists of, roughly, 65 per cent. of moisture, which must of necessity be constantly replaced from external sources.

A pure drinking-water should contain little or no sediment, and any present should be entirely of a mineral character. The microscopic examination of a sample of tap-water which has undergone proper filtration will, in the majority of cases, reveal little or nothing to the observer. But where a water has been improperly purified or stored, many animal and vegetable substances may be present, and in the case of some of these a source of grave danger is indicated. Particularly is this the case where the water is drawn from a well, and in considering such samples notice should also be taken of the situation and

FIG. 96.—FILM OF MILK.

FIG. 97.—ENGLISH DIATOMS.

FIG. 98.—SEWAGE FUNGUS.

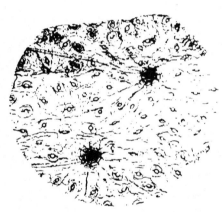

FIG. 99.—LEAF OF TEA.

FIG. 100.—SECTION OF COFFEE-BERRY.

FIG. 101.—TYPICAL CELLS OF COFFEE.

[To face p. 333.

surroundings. For instance, if a manure heap should be found in fairly close proximity to the well, this fact would probably explain any indication of manurial refuse which might be detected in the water, and would warrant the immediate removal of the offending matter.

We will not take into consideration here pond or other open waters, as such sources would, of course, yield unlimited organisms and foreign matter of various kinds.

In conducting a microscopical examination of drinking-water it is advisable to allow about a quart to settle for, say, twenty-four hours, the top of the container being meanwhile covered to exclude dust, etc. At the end of this time the bulk of the water is poured off or siphoned from the top with as little disturbance as possible, after which the remainder is well stirred and conveyed to a conical glass, in which it should be allowed to settle, as before, for about twelve hours. The bulk of the water is again withdrawn, and the sample for micro-inspection taken with a pipette from the bottom, or ' point,' of the conical glass. The maximum of sediment is thus obtained with a minimum of water. The drop is placed on a slip under a thin cover-glass in the usual way.

As before stated, a good quality water will contain little solid matter. Where pollution has taken place, however, the variety of substances which may be met with is enormously wide. To enumerate them would need much space, while to recognize them all at sight would be beyond the power of any microscopist we have so far had the honour of meeting. We must therefore content ourselves with naming a few of the commoner forms met with, and would refer the reader to the various textbooks for a fuller list and more extended descriptions.

1. *Inanimate Matter.*—Mineral substances: Lime, clay, sand, etc., with occasionally traces of lead, zinc, or copper.

Vegetable substances: Starch - grains, ducts, vessels and cuticles of plants; fragments of wood, roots, etc.

Particles of wool, silk, and cotton; human hairs and epithelium; muscle fibres; animal hairs and scales, etc.

2. *Living Organisms.*—Vegetable: Bacteria, diatoms, desmids, Volvox, Protococcus, Chara, Confervæ, Beggiatoa, etc.

Animal: Rotifera, Vorticella, Paramœcium, Amœba, Euglena, tapeworms and their eggs, etc.

Mineral substances are, as a rule, readily recognizable, and have no particular significance. This does not apply to lead, zinc, or copper, either of which may be present in dangerous quantities. The detection of these metals properly belongs to ordinary chemical analysis, as does the estimation of ammonia, nitrates, nitrites, etc.; but a test may with little trouble be made on the stage of the microscope.

Put a few drops of the original water into a cell, and acidulate with a minute quantity of hydrochloric acid. Add a drop or two of potassium ferrocyanide. If iron is present, a blue coloration results; if copper, the water turns bronze; while zinc causes a white precipitate to fall. For lead, test another sample with potassium chromate, which gives a yellow reaction.

Fragments of vegetable origin are usually quite harmless in themselves, but they suggest that the water-supply is open to the atmosphere.

The presence of dead animal matter (fibres, hairs, etc.) must always be looked upon with suspicion. Although a water containing such substances may be in itself quite safe, yet it is evident that serious pollution is at least *possible*, and every effort should be made to trace the source of such a supply.

Of the living organisms, bacteria are the most significant. Where their presence is suspected, it is desirable to make a cultivation from the sediment. This is done in petri dishes on nutrient gelatine, and the colonies are counted and estimations made in the usual manner. Amongst the important organisms to be looked for are *B. coli communis* (the organism normally found in the human intestine, but which becomes virulent in typhoid infection), and *B. typhosus* (typhoid fever). Many other bacteria may be present, but much training and skill is necessary for their detection, and a textbook on bacteriology must be consulted for the details of procedure.

The presence of diatoms and other similar forms of vegetable life does not demand any special notice. While, however, most of these objects are practically without danger, some of them impart certain characteristics to the waters which they inhabit. For instance, according to Whipple, a fishy odour is given by

Volvox (the beautiful globe-like plant often found in pond-water), an aromatic odour by certain diatoms (Meridion, Tabellaria, etc.), and so on.

Fig. 97 shows a number of the commoner forms of British diatoms.

Beggiatoa alba (Fig. 98) is suggestive of sewage contamination, the fungus being found in considerable quantities on sewage farms. It contains sulphur, which accounts for the occasional production in its presence of an odour of sulphuretted hydrogen. Beggiatoa forms masses of long slender threads which are devoid of colouring matter excepting towards the free extremity, where pigmented granules may be seen.

Of the animalculæ to be found in water, many forms will be familiar to the microscopist. The possibility of sewage contamination is suggested by the presence of Amœba, Paramœcium (Infusoria), Vorticella, etc., while the larger organisms, such as the water-fleas, worms, etc., call for no comment here.

If the eggs or parts of Entozoa (tapeworms) be found, the water should be unhesitatingly condemned for drinking purposes, as these eggs are capable of retaining their vitality for considerable periods, and of developing on finding their way to a suitable host—which, in many cases, may be man. (See under ' Flesh Foods.')

Tea.—This is the dried leaf of *Camellia thea*. There are several species which are cultivated by the grower, but the variation in commercial teas depends rather upon the selection of the leaves and the details of preparation than upon any botanical differences. Nearly all teas as sold are ' blended,' various kinds being mixed, so as to secure a more perfect result.

The tea-leaf is best prepared for microscopical examination by soaking in hot water, unrolling, and placing between two slips. As, however, such a leaf is very opaque, it is desirable to employ some means of rendering it transparent, and several ways have been suggested. Dr. Wynter Blyth proposes a simple method as follows : ' A portion of leaf is enclosed between two cover-glasses, and a weight (say a coin) placed on the upper cover. The whole is then heated with a strongly alkaline solution of permanganate of potash. The action commences at once, and the substance must be examined from time to time to see that the oxidation does not proceed too far. The solution attacks the colouring matter and cell contents first, then the cell membranes. When the action is judged to be sufficient—*e.g.*, when the membranes of the leaf only remain—the fragment of leaf is washed in water, and treated with a little strong hydrochloric acid, which dissolves the manganese oxide which has been formed, leaving a translucent white membrane. Tea-leaf thus treated is quite different in appearance from other leaves.'

The same author suggests the preparation of what he terms a ' skeleton ash.' This is obtained by enclosing between two covers as above directed, and burning on a metal sheet the leaf thus arranged. Such preparations may be preserved by cementing the edges of the two covers, or by fusing them in a flame, and will be found to be of the greatest assistance in identification.

The leaf of the tea-plant is elliptical, having its margin serrated almost, but not quite, up to the stalk. The ribs or veins of the leaf run from the midrib to within a short distance of the margin, when they curve inwards, thus leaving a clear space around the edge of the leaf. This clear space is characteristic, not occurring in the sloe, willow, beech, hartshorn, or elder —the leaves most employed when adulteration takes place.

The epidermis should be removed from the under surface of a leaf, and examined separately (Fig. 99). It will be found to contain numerous stomata and hairs. The former are present in good numbers, are oval or round, and are composed of two guard cells. The hairs are simple and pointed, rather numerous on young leaves, but less abundant on old. They are about 1 mm. in length, and 0·015 mm. in breadth. A vertical section of the tea-leaf exhibits very distinctive idioblasts. These occur most frequently in the older leaves, but are never entirely absent. They are long, tough, and branched, and may be shown, without sectioning the leaf, by Moeller's method of warming some fragments of leaf in a very strong solution of caustic potash, and then pressing these firmly under a cover glass. The presence of the idioblasts is strong evidence of the nature of the leaf.

Tea contains an alkaloid principle called ' theine,' which is poisonous when taken in large quantities. Theine may be

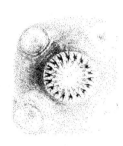

FIG. 102 —HEAD OF CYSTICERCUS.

FIG. 103.—TRICHINA SPIRALIS

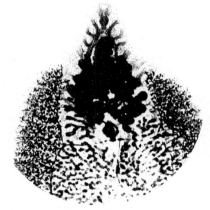

FIG. 104.—OXYURIS VERMICULARIS

FIG. 105.—LIVER FLUKE OF SHEEP.

FIG. 106.—BACILLUS ANTHRACIS.

FIG. 107.—HUMAN BLOOD-DISCS

[To face p. 310.

simply demonstrated by placing a little tea on a watch-glass and covering this with a second glass of the same size. These are then placed on a water-bath, or on a gauze over a Bunsen flame, and in about five minutes drops of moisture will be seen on the upper glass. In about ten minutes the theine, in its characteristic crystallized form of long microscopic needles, may be seen under a low power; while after a further five minutes quantities of the crystals will have been produced. As no theine is volatilized from exhausted leaves, this test is of value in detecting tea which has been redried.

Coffee.—Coffee is the dried seed of *Caffea arabica*, with the husk removed. This seed is roasted, during which process is liberated the oil—caffeol—to which the aroma of coffee is due.

The main portion of the coffee-berry is composed of thick-walled cells (Fig. 100), which, when cut in radial sections, present a distinctive knotted and irregular appearance. A thin, tough membrane covers the berry, and this membrane, fragments of which may be found in ground coffee, contains the typical spindle-shaped cells illustrated (Fig. 101). In the unroasted berry globules of oil may be detected, but, as above stated, these are dissipated on heating.

Coffee is chiefly adulterated by an admixture of either chicory, various cereals, ground date-stones, caramel, etc., the first-named being the most frequently found.

If a small quantity of roasted and ground coffee be sprinkled on the surface of cold water, any chicory present will sink rapidly, while the coffee, by reason of its oily nature, will float for some time. Further, the sediment of chicory is softened, while the pure coffee remains hard. Microscopically, chicory differs widely from pure coffee, the soft tissues of the former abounding in spiral vessels, which are usually readily recognizable. The roasted substance of beans, wheat, acorns, etc., which have been used as adulterants, are quickly detected by the presence of the respective starches, which are coloured by the iodine test. Neither coffee nor chicory give the characteristic blue reaction. The epidermis of date-stones has more or less irregularly shaped oblong cells, unevenly thickened and frequently pitted; while caramel (burnt sugar) may be detected under a low power by its shining particles and its solubility in hot water.

Cocoa.—Commercial cocoa is the seed of *Theobroma cacao*, powdered and subjected to various processes. The original tissues are very characteristic, but the finely powdered product usually offered for sale is not easy to identify microscopically; and, indeed, a supplementary analysis by chemical methods is almost indispensable.

If a little powdered cocoa is stirred in warm water, and a drop placed upon a slide, the starch-grains may be distinguished. These are not unlike rice-starch, but differ essentially in that the rice granules are angular, while those of cocoa are rounded. There are also present small masses of tissue in which are occasional dark brown cells. These contain the pigment known as 'cocoa-red,' and are coloured red by dilute sulphuric acid and violet by a solution of acetic acid and alcohol (Fig. 108).

On heating the cocoa to boiling-point, numbers of oil-globules separate. The starch, however, gelatinizes very slowly, and the iodine reaction is also very slow.

To further examine, remove the fat from some powdered cocoa, by shaking with alcoholic ether at intervals for some hours; wash with alcohol, and examine in water. Add a little chloral iodine and boil gently. The starch is now blue, and any epidermal cells or crystals present are observable.

Cocoa contains an alkaloid termed 'theobromine' which is closely related to caffein and theine. It sublimes at 170° C., forming microscopic needles. The principal adulterants are sugar, starch, and other more harmful substances; but the detection of most of these must be referred to means other than microscopical.

Mustard.—Mustard is the seed of the black and white mustard plants, *Brassica nigra* and *Sinapis alba*. In preparing the condiment as usually sold, the husks of the seeds are removed, and the latter ground to a fine flour, which is separated into various grades by passing it through sieves. The different mustards are not usually sold in a 'pure' state. The commercial product consists of a mixture of the white and black flours, with or without wheat-starch, the larger the proportion of black flour, the stronger and more pungent being the condiment.

White mustard-seed consists of an outer husk and the seed

proper. The husk is of very complicated structure, being made up of six coats or layers, the outer of which is of a mucilaginous nature (Fig. 109). In the husk is also found the aleurone layer —cells containing a proteid substance common to many varieties of seeds. In ground mustard, however, the whole of the husk has usually been removed, and one of the most valuable means of identification is thus lost.

The substance of the seed consists entirely of minute oil-bearing cells. These are very similar in appearance to starch granules, but they do not give the usual iodine reaction, nor do they polarize light. Mustard may be regarded as one of the purest food substances sold. When adulteration takes place, this is usually in the nature of an addition of wheat or other flour. On the application of the iodine test, any such admixture is at once revealed, the starch turning blue. As mustard is naturally entirely free from starch, the reaction above mentioned is positive evidence of an added substance. Any aleurone grains present assume a yellowish-brown tint under this test.

Mustard is sometimes coloured by the addition of turmeric. This is detected by subjecting the sample to the action of ammonia, which produces a brownish-red colour.

Pepper.—Both black and white pepper consist of the fruit or berry of *Piper nigrum*, the difference being that the former is the fruit in an unripe condition, while in the latter it is mature, and deprived of its outer layer. Unlike mustard, pepper contains a quantity of starch, which forms the central mass of the peppercorn. It contains also about 1 per cent. of natural oil.

If a peppercorn be sectioned, it may be easily identified with the diagram given in many textbooks or with a type specimen to be obtained from the usual sources (Fig. 110). The outermost layer consists of hard cells, more or less oval in shape, and having a curious ' pleated ' appearance ; while the other tissues in their sequence are the outer mesocarp, fibrovascular layer, oil-cells, pigment layer, hyaline tissue, and the aleurone starch and resin cells.

A little powdered pepper (Fig. 111) should be diffused in a film of water, when the minute starch-grains may be identified by the iodine test (p. 317). Moisten some pepper with a drop of alcohol, allow it to stand for a few moments, and then add a drop of dilute glycerine. Cover, and after about five minutes the preparation may be examined. A number of long, narrow prismatic crystals may now be seen; these are piperine, one of the alkaloid principles of pepper. An orange-red coloration is imparted by treating with concentrated nitric acid, and if the crystals be further subjected to the action of caustic potash, a blood-red tint is produced.

The adulterants of pepper are legion, although probably not many are now employed. A number of these are not detectable by microscopical means alone, and, in any case, reference should be made to the works on the subject for a full description of minute structure, etc.

Sugar.—The microscopical examination of sugar reveals but little of value to the observer, and it is probable that this is the purest of all ' manufactured ' foodstuffs. If a sample will pass an ordinary physical examination it may be fairly assumed to be fit for consumption. In the case of powdered or granulated sugars, there is, of course, always the possibility of an accidental admixture of dust, débris from packing, etc., and this is best detected by microscopical means. Again, in the sugar known as ' Demerara ' may sometimes be found. the ' sugar-louse,' *Lepisma saccharina*, well known to microscopists by reason of the distinctive markings on its scales, which are frequently used as tests.

Various sugars take characteristic forms on recrystallization, and such preparations are very beautiful and of value in identification (Fig. 112). It is, however, a difficult matter to obtain constant forms, as the crystals vary considerably according to the method of drying, etc.

Flesh Foods.—The microscopy of flesh foods is in a somewhat different category from that of other substances which we have. been considering in these pages. Our efforts hitherto have been mainly directed towards a detection of adulteration; but as this is not possible with flesh foods, at least in the case of fresh meat, we have now to turn our attention to the recognition of diseased or otherwise unhealthy tissues. So far, of course, as freedom from putrefaction goes, our eyes and noses are the best detectives; but we need something more than this for complete

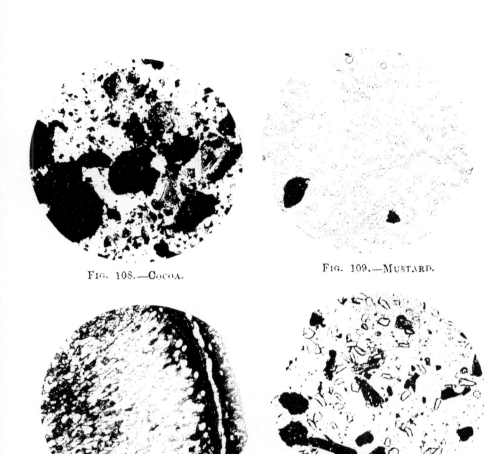

FIG. 108.—COCOA.

FIG. 109.—MUSTARD.

FIG. 110.—SECTION OF PEPPERCORN.

FIG. 111.—PEPPER.

FIG. 112.—CANE-SUGAR (RECRYSTALLIZED).

FIG. 113.—SEGMENT OF TÆNIA SOLIUM

[To face p.

safety, and as a means of discovering more obscure troubles the microscope is invaluable. In fresh meat we have to consider the possibility of several dangers—namely, the presence of animal parasites, Entozoa, etc.; pathological conditions of the flesh itself—tubercle, anthrax, and other diseases due to bacteria; and the less important addition of borax, nitre, or salt, which may have been used as a preservative.

We cannot here extend our inquiry to cover the composition of canned or potted meats, sausages of various kinds, or other forms of meat more or less disguised. The analysis of these is a task much more lengthy and complex, for in such cases we have to identify, perhaps, sundry dyes and adulterants, besides various parts of the animal not usually eaten in a fresh condition.

We will therefore turn our attention first of all to the internal parasites or worms, usually included under the term 'Entozoa.' Amongst the most important of these are—

Cysticercus cellulosæ, the larval form of *Tænia solium*. These are small round or oval cysts, most frequently found in the muscle of the pig, embedded in the connective tissue between the fibres. The cyst has a greyish appearance, and is about the size of a small bean. When the infected animal is killed, and these cysts are transferred in the meat to a suitable host—*e.g.*, man— the larva develops into the tapeworm, *Tænia solium* (Fig. 113). This worm is distinguished by its round head, which is furnished with four suckers and about twenty-six hooklets arranged in the form of a circle (Fig. 102). It may reach a length of 2 metres.

Cysticercus bovis, the cystic form of *Tænia saginata* (= *T. medio-canellata*). *Cysticercus bovis* is found in the muscles of the ox, cow, and calf, from which it is transmitted to man when the animal is killed and eaten as food. It then develops into the tapeworm, and may attain the enormous length of 4 metres. *Tænia saginata* has a large, flattened head, furnished with four suckers, but devoid of hooks.

Tænia echinococcus: A small tapeworm, about 5 mm. in length, developing especially in the lungs and liver of herbivorous animals. The eggs of this parasite are readily developed in man, and may produce fatal results.

Trichina spiralis: This is a minute worm which is found

encysted, most frequently in the muscle fibres of the pig. The male worm is about $1\frac{1}{2}$ mm. in length, the female $3\frac{1}{2}$ to 4 mm., and these are found coiled up within a calcified cyst (Fig. 103). In this condition it is comparatively harmless, and its development only takes place when the host is killed. The cyst is then dissolved by the digestive juices in the new host, when the worms are free to pass into the intestine. Here a more serious, and even fatal, disturbance may be caused in man; but if the infection be slight, the worms may pass to the muscles of the new host, again become encysted, and the person suffer no subsequent discomfort. In examining for trichinæ, etc., sections should be cut in the direction of the muscle fibres, choosing pieces near to attachment to sinew or bone. Prepare as directed by Mr. Cole, and examine with low and medium powers.

Distomum hepaticum: The liver fluke (Fig. 105). A large, flat organism $\frac{1}{4}$ inch to 1 inch in length. It is commonly found in the livers of sheep, but may occur in other animals. In the fully developed stage the fluke is not injurious to man.

Tuberculosis.—This disease affects several of the domestic animals—the cow, and pig, and poultry. The tubercles are found in various stages in the lungs, and later, as the disease gains a firmer hold, in the other organs and the muscular tissues. Sections should be prepared in the usual way, and the bacillus searched for. When the infection is confined to certain organs of the animal, such as the lung and lymphatic glands, it is not usual to condemn the entire carcass; but where the disease has become general and the animal is emaciated, it must not be used for human food.

Other bacteriological infections are septicæmia, swine fever, tetanus ('lockjaw'), cholera, anthrax (Fig. 106), glanders, etc. Where any of these are suspected, pieces of the flesh should be placed in alcohol to harden, and then embedded, cut and stained, the organisms being then identified by reference to works on the subject.

If systematic work is to be done in food-microscopy it is suggested that a set of type specimens of food substances be arranged, to aid in the identification of samples under notice.

Most typical tissues may be obtained ready prepared, while other substances may be mounted by the worker himself as opportunity offers.

A nucleus collection might be on the following lines :

Water.—A full selection of possible objects would be very extensive, and it is advised that such substances as point to organic pollution are the most important for our purpose. Typical fibres of wool, cotton, flax ; epithelial cells ; *Beggiatoa alba ;* a slide of strewn diatoms ; Infusoria (say Paramœcium) ; Amœba.

Milk.—Colostrum bodies ; a preparation containing pus cells, and one of blood-discs.

Cheese.—*Aspergillus glaucus ;* cheese-mites.

Flour.—Starches of wheat, rye, and barley ; aleurone-grains ; wheatgrains.

Tea.—Portion of leaf showing hairs and stomata.

Coffee.—Section of berry ; ground coffee ; chicory.

Cocoa.—Pure cocoa.

Mustard.—Pure mustard.

Pepper.—Ground pepper ; section of peppercorn.

Meat.—Section of voluntary muscle ; fatty (adipose) tissue ; segment of tapeworm (*Tænia solium*) ; head of *T. solium ;* head of cysticercus ; liver fluke ; *Trichina spiralis* in muscle ; tubercle.

The cost of the above slides would not exceed, say, £2 2s., and they would be found of considerable service.

The principal reagents for the work are noted below. Most of these are best purchased ready prepared, but we append formulæ for the convenience of readers :

Reagents for Starch.—Iodine, 2 grammes ; potassium iodide, 1 gramme ; distilled water, 200 grammes. Make a clear solution. Stains starch blue, proteid yellow ; cellulose yellow, the latter turning blue when treated with concentrated sulphuric acid.

Reagent for Proteid (*Millon's*).—Mercury, 3 c.c. ; fuming nitric acid, 27 c.c. Dissolve without heat ; then add an equal volume of water. Gives a brick-red reaction in the presence of proteid.

Chloral Iodine.—Chloral hydrate, 50 grammes ; water, 20 c.c. Dissolve, and then add iodine until the solution is saturated. Keep a few crystals of the iodine in the bottle.

Osmic Acid Solution.—A 1 per cent. aqueous solution of osmic acid. Should be protected from light. Colours fat dark brown or black.

Litmus solution. Sulphuric acid.

Distilled water. Glycerine. Alcohol.

The following are suggested works of reference

' Foods : their Composition and Analysis.'   A. and M. Wynter Blyth.
Microscopical Examination of Foods and Drugs.'   H. G. Greenish.
' Flesh Foods.'   C. Ainsworth Mitchell.
' Practical Sanitary Science.'   D. Sommerville.
' A Compendium of Food-Microscopy.'   E. G. Clayton.
The Microscopy of Drinking Water.'   G. C. Whipple.
The Microscopic Examination of Drinking Water.'   J. D. Macdonald.
' Manual of Bacteriology.'   Muir and Ritchie.

INDEX

A

ABBE'S apertometer, 64
  camera lucida, 122
  illuminator, 94
  test-plate, 75
Aberration, chromatic, 51
  spherical, 53, 76
Absolute alcohol, 142
Acarus domesticus, 304
Accessories, sundry, 133
Acetate of copper screen, 92
  of copper solution, 189
Achromatic *v.* apochromatic objectives, 57
Achromatism, 50
Actinomycosis, 171
Adjusting a microscope, 129
Aleurone, 192
Algæ, green, preserving fluid for, 189
  marine, 192
  staining and mounting, 181
Alum, 303
Ammonium bichromate, 143
Analyzer, 108
Anchylostoma, 173
Andrews, Cuthbert : ' Microscopy of Foods,' 299
Angular apertures, 62
Aniline blue-black, 156
  blue, 156
Animal tissues, hardening and preserving, 142
Annular vessels, 180
Antheridia of fucus, 191
  of mosses, 190
Anthers of flowers, mounting, 177
Apertometer, Abbe's, 64
  R. and J. Beck's, 66
  Cheshire's, 66
Apertometers, 64-66
Aperture, numerical, 63
Apertures of objectives, 62
Aplanatic aperture, how ascertained, 97
  of condensers, 95
  of sub-stage condenser, 97

Aplanatic magnifiers, 36
Aplanatism, 51
Apochromatic correction, 51
  v. achromatic objectives, 57
Aquaria for Rotifera, 250
Aquarium microscope, 252
Aqueous media, 162
Archegonia of mosses, 190
Arragonite, interference figures of, 236
Ashe's camera lucida, 122
  two-speed fine adjustment, 26
Aspergillus glaucus, 304
Attachable mechanical stages, 13

B

Bacillus, anthrax, 169
  of enteric fever, 171
Bacillus tuberculosis, staining and mounting, 168
Back lens of objective, 102
Bacteria, examination of living, 106
  preparation of, 167
Baker's D.P.H. microscope, 9
Barley, mounting, 177
  starch, 302
Bath, embedding, 147
Bausch and Lomb objectives, 79
  research microscope, BB, 17
Beale's camera lucida, 122
Beck, R. and J.'s, apertometer, 66
  condensers, 95
  Imperial microscope, 25
  selenites, 109
  two-speed fine adjustment, 26
Beggiatoa alba, 309
Berlin blue watery solution, 175
Bertrand plate, 228
Bertrand's quarter-quartz plate, 233
Biaxial crystals, 224
Binocular eyepieces, 87
  microscope for Rotifera, 256
  Greenough, 35, 39
  Stephenson, 35
  microscopes, 33